T0313349

Engineering Education
for Sustainability

RIVER PUBLISHERS SERIES IN MANAGEMENT SCIENCES AND ENGINEERING

Series Editors:

J. PAULO DAVIM
University of Aveiro
Portugal

CAROLINA MACHADO
University of Minho
Portugal

Indexing: All books published in this series are submitted to the Web of Science Book Citation Index (BkCI), to SCOPUS, to CrossRef and to Google Scholar for evaluation and indexing.

The "River Publishers Series in Management Sciences and Engineering" looks to publish high quality books on management sciences and engineering. Providing discussion and the exchange of information on principles, strategies, models, techniques, methodologies and applications of management sciences and engineering in the field of industry, commerce and services, it aims to communicate the latest developments and thinking on the management subject world-wide. It seeks to link management sciences and engineering disciplines to promote sustainable development, highlighting cultural and geographic diversity in studies of human resource management and engineering and uses that have a special impact on organizational communications, change processes and work practices, reflecting the diversity of societal and infrastructural conditions.

The main aim of this book series is to provide channel of communication to disseminate knowledge between academics/researchers and managers. This series can serve as a useful reference for academics, researchers, managers, engineers, andother professionals in related matters with management sciences and engineering.

Books published in the series include research monographs, edited volumes, handbooks and text books. The books provide professionals, researchers, educators, and advanced students in the field with an invaluable insight into the latest research and developments.

Topics covered in the series include, but are by no means restricted to the following:

- Human Resources Management
- Culture and Organisational Behaviour
- Higher Education for Sustainability
- SME Management
- Strategic Management
- Entrepreneurship and Business Strategy
- Interdisciplinary Management
- Management and Engineering Education
- Knowledge Management
- Operations Strategy and Planning
- Sustainable Management and Engineering
- Production and Industrial Engineering
- Materials and Manufacturing Processes
- Manufacturing Engineering
- Interdisciplinary Engineering

For a list of other books in this series, visit www.riverpublishers.com

Engineering Education for Sustainability

Editor

João Paulo Davim

University of Aveiro
Portugal

Routledge
Taylor & Francis Group

LONDON AND NEW YORK

Published 2019 by River Publishers
River Publishers
Alsbjergvej 10, 9260 Gistrup, Denmark
www.riverpublishers.com

Distributed exclusively by Routledge
4 Park Square, Milton Park, Abingdon, Oxon OX14 4RN
605 Third Avenue, New York, NY 10017, USA

Engineering Education for Sustainability / by João Paulo Davim.

© 2019 River Publishers. All rights reserved. No part of this publication may be reproduced, stored in a retrieval systems, or transmitted in any form or by any means, mechanical, photocopying, recording or otherwise, without prior written permission of the publishers.

Routledge is an imprint of the Taylor & Francis Group, an informa business

ISBN 978-87-70221-04-7 (print)

While every effort is made to provide dependable information, the publisher, authors, and editors cannot be held responsible for any errors or omissions.

Contents

Preface

Currently, sustainability is a key concept in engineering education. One of the most widely recognized definitions is based on Brundtland report, *"development that meets the needs of the present without compromising the ability of future generations to meet their own needs"*. Being understood as a key issue in the current societies, sustainability is characterized by its three essential pillars, namely, environment, society, and economy. Engineering education field is not an exception. Indeed, presently, the integration of sustainability in engineering education is a relatively recent phenomenon and a great challenge, presenting the information about engineering education for sustainability a great interest to improve communication between professors, researchers, and students in universities, institutes, and research laboratories.

Designed to increase the knowledge of all those that are interested in develop in engineering education the sustainability concept and ideas, this book, entitled *Engineering Education for Sustainability* is organized into five chapters. Chapter 1 covers *"Experiences from 5 years of educating sustainability to computer science students"*. Chapter 2 discusses *"Review of Decision Support Methods in Green and Sustainable Supply Chains"*. Chapter 3 describes *"Analyzing the drivers of engineering education for sustainability using MCDM approach"*. Chapter 4 discusses *"Visualization technologies in construction education: A comprehensive review of recent advances"*. Finally, in chapter 5, *"A Legal Framework and Compliance with Construction Safety Laws and Regulations in Vietnam"* is presented.

The interest in this book is evident for many types of institutions, namely, research laboratories, institutes, and universities all over the world. The Editor acknowledges to RIVER Publishers for this chance and for their professional support. Finally, I would like to thank to all chapter authors for their availability to work on this editorial project.

J. Paulo Davim
Aveiro, PORTUGAL

List of Contributors

Akeem Pedro, *Department of Architectural Engineering, Chung-Ang University, Seoul 06974, South Korea*

Anh-Tuan Pham-Hang, *School of Computer Science and Engineering, International University – Vietnam National University HCMC; Ho Chi Minh City 7000000, Vietnam*

Birgit Penzenstadler, *California State University Long Beach, Long Beach, USA*

Carlos Francisco Simões Gomeš, *Fluminense Federal University, Brazil*

Davidson de Almeida Santos, *Fluminense Federal University, Brazil*

Eric Rondeau, *University of Lorraine, Nancy, France*

Hai Chien Pham, *Faculty of Civil Engineering, Ton Duc Thang University, Ho Chi Minh City 7000000, Vietnam*

Jari Porras, *Lappeenranta University of Technology, Lappeenranta, Finland*

Karl Andersson, *Luleå University of Technology, Luleå, Sweden*

Osvaldo Luiz Gonçalves Quelhas, *Fluminense Federal University, Brazil*

Quang-Vu Pham, *Faculty of Construction, Ho Chi Minh City of Transport and Communications, Ho Chi Minh City 7000000, Vietnam*

Rahat Hussain, *Department of Architectural Engineering, Chung-Ang University, Seoul 06974, South Korea*

Rodrigo Goyannes Gusmão Caiado, *Fluminense Federal University, Brazil*

Rohit Agrawal, *Department of Production Engineering, National Institute of Technology, Tiruchirappalli-620015, Tamil Nadu, India*

S. Vinodh, *Department of Production Engineering, National Institute of Technology, Tiruchirappalli-620015, Tamil Nadu, India*

Sheila da Silva Carvalho Santos, *Celso Suckow da Fonseca Federal Center for Technological Education, Brazil*

Thi-Thanh-Mai Pham, *Faculty of International Trade, College of Foreign Economic Relation, Ho Chi Minh City 7000000, Vietnam*

Victoria Maria Palacin Silva, *Lappeenranta University of Technology, Lappeenranta, Finland*

List of Figures

List of Tables

List of Abbreviations

AEC	Architecture Engineering and Construction
AHP	Analytic Hierarchy Process
AID-DS	Analysis of Impacts During the Design for Sustainability
AR	Augmented Reality
BIM	Building Information Modeling
BiM	Building Interactive Modeling
C2C	Cradle to Cradle
CAD	Computer Aided Design
CAVEs	CAVE Automatic Virtual Environments
CRAN	Research Center for Automatic Control in Nancy
CSLF	Construction Safety Legal Framework
CSR	Corporate Social Responsibility
DWS	Department of Work Safety
EIs	Educational Institutions
ELECTRE	ELimination Et Choix Traduisant la REalité
G/S-SCM	Green and Sustainable Supply Chain Management
GDP	Gross Domestic Product
GeSI	Global eSustainability Initiative
HCI	Human-Computer Interaction
HMDs	Head Mounted Displays
IBAM	Interactive Building Anatomy Modeling
ICT	Information and Communication Technologies
IFC	Industry Foundation Classes
ITMO University	St. Petersburg National Research University of Information Technologies, Mechanics and Optics
MBSE	Model-Based Systems-Engineering
MCDM	Multi-Criteria Decision Making
Mgsn	Mobile-GSN
MOC	Ministry of Construction
MOET	Ministry of Education and Training

MOH	Ministry of Health
MOLISA	Ministry of Labour, Invalids and Social Affairs
MOST	Ministry of Science and Technology
OSHA	Occupational Safety and Health Administration
PERCCOM	Pervasive Computing and Communications for Sustainable Development
PHG	Phu Hung Gia Construction and Investment JSC
PPE	Personal Protection Equipment
PROMETHEE	Preference Ranking Organization Method for Enrichment Evaluations
QiS	Quality in Sustainability
QoE	Quality of Experience
QoS	Quality of Service
QUESTE-SI	Quality System of European Scientific and Technical Education for Sustainable Industry
SLA	Service Level Agreement
SLR	Systematic Literature Review
SME	Most Small and Medium Construction Enterprises
SWEBOK	Software Engineering Body of Knowledge
TBL	Triple Bottom Line
VR	Virtual Reality

1

Experiences from Five Years of Educating Sustainability to Computer Science Students

Jari Porras[1*], Eric Rondeau[2], Karl Andersson[3], Victoria Maria Palacin Silva[1] and Birgit Penzenstadler[4]

[1]Lappeenranta University of Technology, Lappeenranta, Finland
[2]University of Lorraine, Nancy, France
[3]Luleå University of Technology, Luleå, Sweden
[4]California State University Long Beach, Long Beach, USA
E-mail: Jari.Porras@lut.fi; eric.rondeau@univ-lorraine.fr;
karl.andersson@ltu.se; Maria.Palacin.Silva@lut.fi;
bpenzens@gmail.com
*Corresponding Author

Sustainability and sustainable development are emerging trends all over the world. The need for changes is evident and immediate. To tackle all the technical and social sustainability challenges, people need to be educated. To date, computer science and software engineering education has not fully answered to this need. This chapter presents a sustainability-focused ICT programme that builds on top of common SE curricula like SWEBOK and sustainability frameworks by Cai, Rusinko, and Mann. We observe across the past five years how the thesis topics as well as the prevalent industry collaborations have slightly shifted but largely covered all pillars of sustainability. Students are well placed in the job market after graduation and report positive feedback.

1.1 Introduction

"The Earth's climate has changed throughout history" Nasa[1]

Climate change is one of the reasons why sustainability and sustainable development have been discussed for decades. Though lately, due to the ever increasing number of alarming news, discussions have increased and shifted from why to how questions. Johan Rockström had already presented in 2009 in Nature article (Rockström et al., 2009) the nine planetary boundaries (climate change, rate of biodiversity loss, nitrogen cycle, phosphorus cycle, stratospheric ozone depletion, ocean acidification, global freshwater use, land use, atmospheric aerosol loading, and chemical pollution) that are necessary for the life in the Earth. Three of these had already exceeded at the time of the publication of the article. Biodiversity loss[2], boundary with the worst state already back then, has continued to worsen. The latest Living planet report from WWF[3] states that 60% of world's wildlife has wiped out since 1970s and this will have a clear impact to our living ecosystem. Climate change, another boundary breaker, has been discussed much more than the biodiversity loss, although both are essential for the planet to flourish. The UN-based climate change conferences have been held since 1995. Although some people relate the climate agreements to the Kyoto Protocol signed in 1997 or even the earlier earth summit held in Rio in 1992, the real public climate and sustainable development debate started in Paris Climate conference 2015 (COP21). Paris meeting, for the first time, brought all nations into a common cause to undertake ambitious efforts to combat climate change and adapt to its effects, with enhanced support to assist developing countries to do so. The main aim of the Paris Agreement was to strengthen the global response to the threat of climate change by keeping a global temperature rise this century well below 2 degrees Celsius above pre-industrial levels and to pursue efforts to limit the temperature increase even further to 1.5 degrees Celsius[4]. The subsequent climate change conference in Marrakesh (COP22), Bonn (COP23) and Katowice (COP24)[5] have been discussing

---

[1] https://climate.nasa.gov/evidence/

[2] https://www.theguardian.com/news/2018/mar/12/what-is-biodiversity-and-why-does-it-matter-to-us

[3] https://wwf.panda.org/knowledge_hub/all_publications/living_planet_report_2018/

[4] https://unfccc.int/process-and-meetings/the-paris-agreement/the-paris-agreement

[5] https://cop24.gov.pl/key-messages/

of the means to achieve the common goals. The latest meeting in Poland in 2018 emphasized role of technologies and humans in supporting the nature to recover. Although the actions done since Paris meeting have not yet impacted there exists examples that actions can change the course of our planet. For example, CFCs, ingredients causing ozone depletion, were banned 1960s and scientists have said that ozone layer is on track to be fully healed within 50 years[6].

Sustainability and sustainable development are not easily achieved. Many challenges exist. Perhaps the most known definition for sustainable development given by the so-called Brundtland report (Brundtland et al., 1987), "a development that meets the needs of the present without compromising the ability of future generations to meet their own needs", shows some of these challenges. Firstly, sustainability is a long-term effort. What we do now will affect our children and generations to follow. However, it is not good to anticipate future consequences as we live in the present. Secondly, sustainability needs a balance between resources and needs. Unless we can increase the amount of resources, the needs cannot increase. This comes tricky when we look at the non-renewable resources. How much of those we can use without compromising the abilities of the future generations? This is a hard question in the current economy-driven world. This is closely linked to the sustainability dimensions (ecological, economic, and social) presented by the Brundtland report. Ecological dimension has been discussed quite actively but, in many countries, this is related to the green parties or movements. Economical dimension is adopted by businesses to justify their growth economy. The social dimension is far less discussed as well as the combination of all these dimensions. In the end, full sustainability would include all of these dimensions (Brundtland et al., 1987).

Information and communication technologies (ICT), including software solutions have been reported to have both positive (ICT as a part of the solution) as well as negative effects (ICT as a part of the problem) on sustainable development (see Figure 1.1). Global esustainability Initiative (GeSI) has published a series of reports[7] on the effects of ICT to the 21st century challenges on sustainability. Although ICT is part of the problem the benefits of using ICT outperform the drawbacks many times (see Figure 1.2).

---

[6]https://www.nationalgeographic.com/environment/global-warming/ozone-depletion/
[7]http://smarter2030.gesi.org/

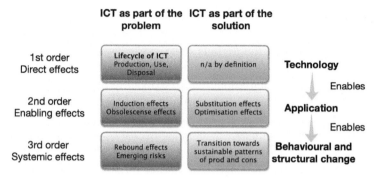

Figure 1.1   Impact of ICT on the sustainable development (Hilty and Aebischer, 2014).

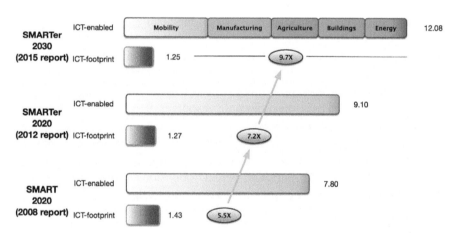

Figure 1.2   Development of ICT impact on the sustainable development (GeSI[7]).

The GeSI reports have illustrated how the ICT footprint has decreased over the years, regardless of the ever-increasing usage, while the ICT-enabled benefits have increased with the new ways of using ICT-solutions. In addition, the benefits reported by the GeSI reports represent only part of the solutions. GeSI reports emphasize only the applications level of using ICT, i.e. enabling effects in Figure 1.1, while the systemic effects of ICT can outperform even those.

ICT has an important role in various application domains (e.g. agriculture, e-commerce, energy production and use, e-working, etc.) by having optimization or substitution effects, as presented in Figure 1.1. Although Figure 1.1 does not recognize direct positive effects of ICT, as ICT will always generate emissions, green IT has been seen as a positive approach to improve the efficiency of ICT. This impacts the decrease in ICT footprint as is thus seen important. The systemic effects would require some behavioural or structural changes that in the end would lead to transition towards sustainable production and consumption and as such would reflect towards the Brundtland definition of sustainable development. All this will however require individuals, societies, and business to have proper knowledge and tools.

Education has an important role in increasing the knowledge and skills of people to whatever they want to achieve. This is well stated by Joseph Tainter in his article (Tainter, 2006) "People sustain what they value, which can only be derived from what they know". In sense of sustainability, Joseph Tainter also presents four questions (a) what to sustain?, (b) for whom to sustain?, (c) for how long to sustain?, and (d) at what cost to sustain?, that in the end matter if we think about the sustainable development. If considering the climate change and our planet, these questions should be easy to answer. But do we have the knowledge or means for the required change?

Sustainable development has been covered by many fields and there exists various programmes in environmental engineering, chemistry, marine science, etc. that emphasize at least some aspects of sustainable development. In computer science, these kinds of programmes did not exist for multiple reasons. The ACM/IEEE curricula guidelines[8] for (*Computer Science Curricula 2013 Curriculum Guidelines for Undergraduate Degree Programs in Computer Science,* 2013) or (*Software Engineering 2014 Curriculum Guidelines for Undergraduate Degree Programs in Software Engineering,* 2014) do not really emphasize sustainability in topics. The Software Engineering curriculum guidelines mention sustainability only once that is in the example of SE program of Mississippi State University. The newest version of CS curriculum guidelines (since 2013) has included sustainability in two instances. Sustainability is covered in two hours in Social issues and professional practice knowledge area as well as an elective in Design-oriented HCI under the HCI (Human-Computer Interaction) knowledge area. One should, however, notice that both of these curricula guidelines are 4 to 5 years old and it is expected that the sustainable development will be covered in the new versions on some level (these guidelines have traditionally been following various bigger trends quite closely). The Software Engineering Body of Knowledge

---

[8]https://www.acm.org/education/curricula-recommendations

(SWEBOK)[9], the guidebook for software engineering discipline, is no better as it mentions sustainability once under economics. This gives the impression that sustainable development under software engineering or computer science has not been seen too important.

This book chapter aims to present one approach to teach sustainability to computer science students. This approach was implemented as European Erasmus Mundus programme and targeted students from various engineering backgrounds but with the idea of linking sustainability to computer science education. Section 1.2 presents the background and fundamentals on educating sustainability in various forms. Section 1.3 presents the program and its contents and in Section 1.4, we look at the outcomes of the program after having 4 cohorts of students graduated and the last cohort starting their last semester. Section 1.5 will conclude this book chapter.

1.2 Related Work

This section explores the connection of sustainability education and computer science education and describes related work as well as work we have built our framework on.

Undergraduate computing education often fails to address our social and environmental responsibility (Mann et al., 2008). Even though Cai (Cai, 2010) and Penzenstadler and Fleischmann (Penzenstadler and Fleischmann, 2011) proposed opportunities for integrating sustainability into undergraduate computing education, computing education has been slow to act towards a shift in adopting sustainability education. The consequence is that there is a deficit in knowledge in how software engineering theory and practice relates to sustainability (Gibson and others, 2017). Even though the higher education sector addresses sustainability at an operational level (Horhota and others, 2014; Tangwanichagapong and others, 2017; Weenen, 2000), few have addressed education for sustainability in a holistic, multidisciplinary, and systematic manner (Mann, 2016).

Amongst the barriers and challenges to the integration of sustainability into the computing education curriculum are a fundamental lack of interest; staff training; a lack of tradition; and a lack of priority; colleagues' scepticism; students' expectations of the course; content vs the actuality of an "expanded" course; an absence of policy; syllabus constraints; lack of

---

[9]https://www.computer.org/web/swebok/index

leadership; an unfavourable view of the role of education for sustainability; and the siloing within faculties of education (Cai, 2010; Down, 2006; Falkenberg and Babiuk, 2014). Consequently, there is a need for establishing an educational framework for sustainability, which requires joint leadership across institutions.

The contribution by Sam Mann below explains how sustainability can be generally integrated into computing and computing education (Mann et al., 2010), which could be done either by a centralized, a distributed, or a blended approach. He proposes to weave the philosophy and the practices of sustainability throughout computing education as opposed to adding on separate modules, see Figure 1.3.

Rusinko (2010) takes this a step further and proposes several options to integrate sustainability into a computing curriculum. She proposes a framework in the form of a generic matrix (see Figure 1.4) of options for integrating sustainability in higher education. It provides a broad, non-discipline-specific orientation; and it is supposed to be applicable at course, program, and cross-disciplinary/cross-university levels. The matrix is flexible in that users can move from one quadrant to another, and can select one or multiple quadrants (or options) with respect to integrating sustainability in higher education. In addition, users can start at whichever quadrant is most comfortable and appropriate for them with respect to integrating sustainability into their curricula (Rusinko, 2010, p. 253).

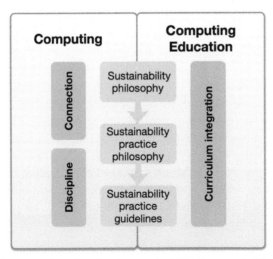

Figure 1.3 The computing education for sustainability framework (Mann et al., 2010).

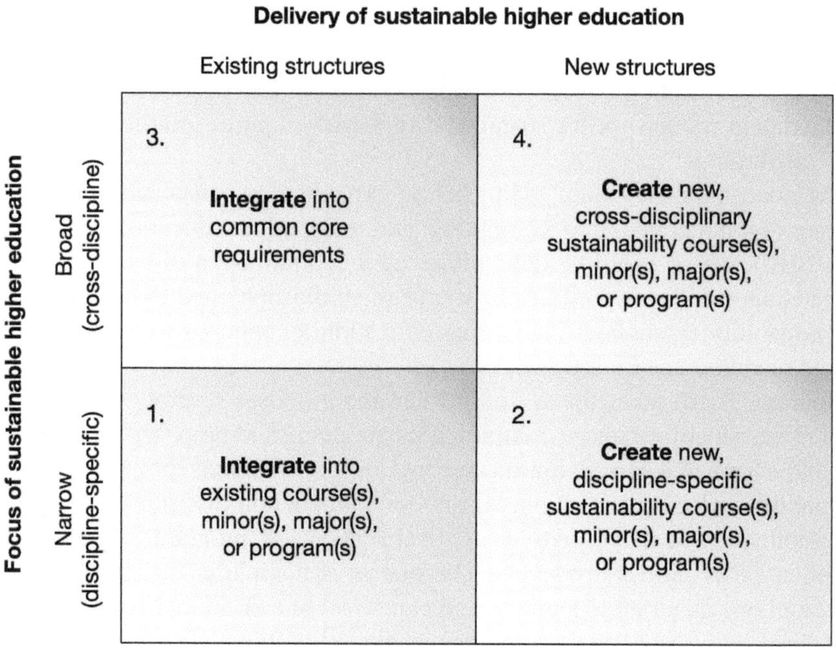

Delivery of sustainable higher education

	Existing structures	New structures

Figure 1.4   Sustainable higher education integration matrix (Rusinko, 2010).

Özkan and Mishra (2015) lay emphasis on adding systems thinking as well as ethics into the curriculum for ICT for sustainability. Integrating systems thinking concepts have also proven helpful in courses on Software Engineering for Sustainability (Penzenstadler et al., 2018).

Samuel Mann proposes to educate every student as a sustainable practitioner in his 2011 book (Mann, 2011). He takes the ideas of all the frameworks proposed above to a new level by instigating that being a sustainable practitioner should be a first-class programme objective as opposed to just one more learning goal in a specific major. He challenges to think about everything with sustainability and long-term consequences in mind.

For PERCCOM, we take on Mann's mindset of educating every student to be a sustainable practitioner, and we add structure the programme inspired by the frameworks of Rusinko and Cai (see Figure 1.5). The programme is described in the following section.

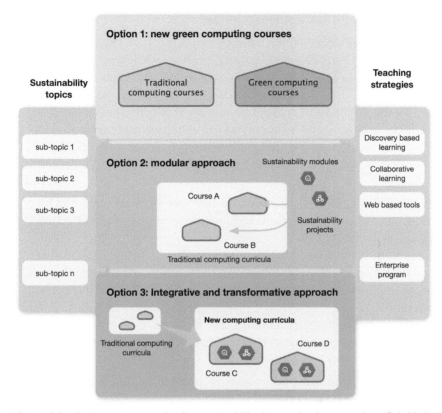

Figure 1.5   Options and strategies for sustainability integration in computing (Cai, 2010).

1.3 PERCCOM Programme

PERCCOM – Pervasive computing and communications for sustainable development is an Erasmus mundus master's programme (2013–2019) on sustainable development. The objective of this programme is to fill the gap between sustainable development challenges and education activities, especially on ICT sector. PERCCOM aims to transfer the emerging sustainability challenges of the businesses and society into educational activities with the emphasis on ICT as a main driver. PERCCOM is a multi-perspective approach as it aims to combine the strengths, competences and views of experts in different ICT domains.

1.3.1 Starting Point for the Programme

PERCCOM programme was applied three times before it was finally accepted by the European Commission. The very first proposal was tied closely to the technical aspects, i.e. computing and communications, of sustainable development. The subsequent proposals broadened the perspective by adding a bit more social and application aspects to the programme structure. Sustainability itself was understood mostly from the environmental perspective. This is partly due to the hosting partners of the project (below) and partly because of sustainability trends at the time of the application process (2010–2012).

- The project coordinator, University of Lorraine, Nancy, France had long history in engineering education and research on networks and network protocols.
- The second host, Lappeenranta University of Technology, Finland, had a long history of combining technology and business and a shorter history of applying sustainable development principles in all of the represented disciplines.
- Luleå University of Technology is also an engineering-oriented university focusing its research and teaching efforts to pervasive systems combining networks and software skill

PERCCOM programme was created to address the following specific challenges (from our original project proposal):

- To understand the emerging sustainability challenges within society and businesses and to transfer them into educational solutions with ICT as a key element,
- To combine the strengths, competences, and experience of experts in different ICT perspectives (e.g. elements of systems as well as whole systems, hardware as well as software, communication and computation, a single phase of an element as well as the whole lifecycle) and thereby develop a common platform of competence within the guidelines of the Bologna process,
- To propose the new International Master degree with no currently available match at international level filling the gap between ICT skills and environmental considerations,
- To attract highly motivated international students to take this challenge and to create new solutions,

- To provide the prospective students with knowledge, skills, and finally competencies in sustainability and ICT to enable a true impact on ecological, economic and social aspects of sustainability, and
- Finally, to fulfil the needs presented by those various reports on ICT's role as a solution or a part of it.

1.3.2 Structure and the Contents

PERCCOM programme (Klimova et al., 2016) was built to be a two-year multi-national programme that is arranged in four separate semesters. Three semesters are used for education and one semester is dedicated for the master's thesis work. Students start their studies in Nancy, France in their first fall. The first spring semester will be taught collaboratively in Lappeenranta, Finland and St. Petersburg, Russia and the second year, i.e. third semester, starts in Skellefteå, Sweden. The fourth and last semester is reserved for thesis work in any of the programme partners, not only in hosting partners. Figure 1.1 shows the general structure of the PERCCOM programme.

The content of the PERCCOM programme was roughly designed at the project application. The contents were divided in four complementing parts, i.e. Technical courses, Industry-based courses, Cultural courses, and Student project/thesis work, as presented in Figure 1.6. Each educational semester was supposed implement these elements. The sustainability aspects were covered in every course rather than having separate sustainability courses. This resembles the modular approach presented by (Cai, 2010). One of the aims of the PERCCOM programme is the ability of the participating universities to move gradually into the integrative and transformative approach of (Cai, 2010). Semesters were further divided content-wise into different high-level themes and after five years of implementation we ended up having the following course modules (this list represents the latest edition of courses and while some courses have been changed the general structure remains the same as presented in (Klimova et al., 2016)).

Semester 1 - Sustainable Computer Network Engineering

Each selected cohort of students will start its studies in Nancy, France, with the computer network engineering focused semester. The objective of this semester is to provide students with fundamental competencies in computer networks and systems engineering in a sustainable way. This semester focuses on green IT perspectives, i.e. how to make the IT solutions more efficient.

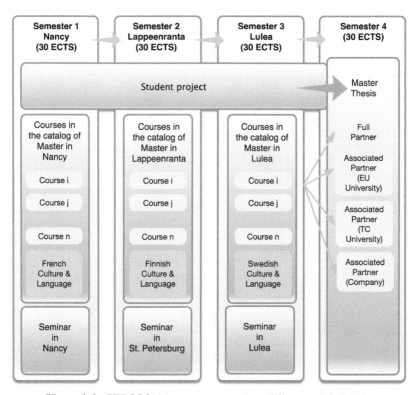

Figure 1.6　PERCCOM programme structure (Klimova et al., 2016).

With the content remaining consistent, the main evolution in this semester has been to improve student project management with regard to the emphasis of sustainability aspects and research method skills.

- Communication protocols (6 ECTS) – ICT and Sustainability
  This lecture focuses on networks fundamentals. It describes the standardization process, the major network architectures, and the main concepts such as switching, encapsulation, errors recovery, connection principles, etc. Classical protocols are presented and detailed. Sometimes, their drawbacks in terms of energy consumption, resources utilization, etc. are highlighted, and some environmental-aware solutions are described.
- Quality of Sustainable Service (6 ECTS) – ICT and Sustainability
  The course provides knowledge on Quality of Service (QoS), Quality of Experience (QoE) and Quality in Sustainability (QiS) in communication

network domain. It gives theoretical and practical knowledge on how to maintain or improve QoS/QiS level at lower layers of the network in terms of delay, reliability, security, energy consumption, recycling, etc.

- Automatic Control for Sustainable Development (3 ECTS) – ICT and Sustainability
  The Research Center for Automatic Control in Nancy (CRAN) is participating and managing many research projects in the environmental domain. The objective of this course is to present to students an umbrella of results on how using automatic control in the context of sustainable development.

- Systems Engineering (3 ECTS) – ICT, Sustainability and Green ICT
  Model-Based Systems-Engineering (MBSE) is presented as a key iterative, collaborative and multidisciplinary approach to define, to develop and to deploy systems in general. MBSE is then applied to ICT domain to enhance a better interoperation between system components, humans as well as technologies, and then to contribute to meet eco-efficiency for C2C (Cradle to Cradle) issues beyond conventional Business and PLM efficiency.

- Sustainable development & circular economy (3 ECTS) – Sustainability
  This course provides an introduction to the general and sectoral regulations of the European Environmental legislation with respect to selected matters of the economic legislation. Competences, principles, and instruments based on the Lisbon treaties, problems of horizontal secondary legislation especially concerning the access to information, process participation, remedy, EIA and environmental liability as well as selected topics of the sectoral plant and product related environmental legislation.

- Specification definition of Master thesis project (3 ECTS) – ICT and Sustainability
  The students start their master thesis from semester 1. All the topics are related to Green IT and sustainable development.

- French Culture and Language (3 ECTS) – Sustainability (social aspects)
  The students start the semester 1 with intensive courses in French to help them in their daily-life in France. Different cultural events are also organized enabling students to discover French art, French history, French food culture, etc.

After the semester 1 the students are expected in sense of sustainable development to be able

- to utilize (new) green ICT metrics to evaluate a level of sustainability of ICT systems and to formalise new green SLA (Service Level Agreement) between ICT companies and their customers. Original approach based on Biomimicry is explained to students to show how to imitate the nature to develop ICT solutions in the context of sustainable development;
- to utilize (new) standards and tools for measuring energy consumed by ICT systems. For example, the students use Cisco Energywise protocol to monitor energy consumed by network architectures, GreenSpector tool to measure energy consumed by software. It conducts the students to elaborate original strategies to efficiently design networked software applications;
- to be systemic when developing Green ICT applications in considering all the stakeholder requirements, the whole life cycle of ICT systems and all the facets of sustainable development (and not only energy). For that, study cases are analysed using system engineering methods to eco-design ICT systems;
- to have expertise in scientific methods for creating new mathematical models on energy consumed by ICT systems. For example, the students use design of experiment for modelling energy consumed by switches.

Semester 2 - Smart software and services

The second semester was arranged in Lappeenranta, Finland, and in St. Petersburg, Russia, and the main aim was to focus on ICT for sustainability, i.e. how ICT can be used in different domains for reducing environmental effects. Lappeenranta was responsible for the technical courses (from the software perspective), cultural courses and student projects, while in St. Petersburg the aim of the courses were to connect students with the software industry and to focus on computing related topics. The content of the technical courses in Lappeenranta has followed the general idea of software development with/for the users with a practical perspective, while in St. Petersburg, the courses were more computing oriented (focus on cloud computing, IoT, blockhains, etc.)

- Green IT and Sustainable Computing (6 ECTS) – Sustainability and ICT for Greening
  This course is the fundamental course of second semester and it introduces the ICT for sustainability approach for the students. This is the only course in second semester that completely focuses on sustainability issues. The sustainability issues are tackled with changing set

of books and scientific articles focusing on various aspects of software (or in general ICT) solutions for sustainability. This course follows the flipped-classroom principles and aims to increase the critical thinking and argumentation skills as well as reflection competencies.

- Home Automation Code camp (4 ECTS) – ICT and ICT for Greening
  Code camp is a highly practically oriented course on software development. Although this course is provided 3–4 times per year with variable content, the topic for PERCCOM students is always related to the sustainable development. During spring 2018, the PER-CCOM students implemented home automation solutions with the aim of intelligently decreasing the energy usage of households.
- Running a software project (6 ECTS) – ICT and ICT for Greening
  Running a software project is a practical software project prom the requirements specification with the real customers through software development all the way to software release. The topics of the team projects are somehow tied to the sustainable topics and aims to increase both technical software development skills as well as soft skills e.g. teamwork and presentation. This course is closely tied with the User and design research course as software development needs to take the users (and customers) into account.
- User and design research in software engineering (6 ECTS) – ICT and Sustainability (social aspects)
  This course emphasizes the role of users in software engineering as well as tools and methods to design for the user. This course follows a project-based approach emphasizing the social and individual aspects of sustainability.
- Transformation of a modern industrial society: The Finnish Model (2 ECTS) – Sustainability (social aspects)
  This course represents the cultural content of the second semester. It was seen more important to teach the students the culture of Finland and Finnish work markets than language as such. This gives the students a cultural perspective to the social sustainability in Finland.
- ITMO – Seminar 1 & 2 (6 ECTS) – Green ICT
  The industry seminars in second semester are arranged together with the St. Petersburg National Research University of Information Technologies, Mechanics and Optics (ITMO University) in St. Petersburg. These seminars have emphasized various technical topics like cloud computing, Internet of Things etc. as well as cultural issues in Russia.

After the semester 2 the students are expected in sense of sustainable development to

- understand the meaning of software in sustainable development in various application domains.
- understand the social (human) perspective and its impact on sustainable development. Software solutions give us tools for affecting on the human behaviour, if the user is properly considered
- be able to implement software solutions for various problems by considering the metrics and methods for sustainable development.

Semester 3 - Smart systems

The aim of the third semester is to focus on wireless networking and systems perspective as well as combining networking and software into efficient pervasive systems. Courses have included topics as network programming, distributed applications, and advanced wireless networks, and lately a course on cloud services was added considering the new local economic activities around data centres in North Sweden (like Facebook, Hydro66, etc.). Moreover, a cultural course including a short introduction to the Swedish language and the Swedish society is included. A code lab or hackathon on green challenges is also being held during this semester. Some courses have been changed but in general the topics have remained the same.

- Network programming and distributed applications (7.5 ECTS) – ICT, ICT for Greening
  The learning outcome of this course is the scientific foundation of network programming and distributed applications including security considerations and the proven experience programmers in this field of Computer Science. Furthermore, students are given the capacity for carrying out teamwork and collaboration with various constellations, both in groups where the students choose whom to work with and in groups put together by others. The goal is also that students can create, analyse and critically evaluate various technical solutions in terms of the design and implementation of communicating computer programs and to show insight in research and development by understanding limitations and possibilities. Also, they should be able to plan and use appropriate methods to undertake advanced programming tasks within predetermined parameters and show the ability to identify knowledge gaps and bridging these gaps by gaining new knowledge. Lastly, but not least,

they should gain the ability to understand, interpret and present scientific publications in the area. Teaching methods include traditional lectures by academic staff members, guest lectures by industry representatives, lab assignments, and seminar presentations by students. The course is finished with a home exam.

- Advanced Wireless Networks (7.5 ECTS) – ICT and Green ICT

  The learning outcome of this course covers details of current radio transmission technologies on the physical layer. Students study how to compute the parameters of the radio transmission zone, to describe the problem of hidden/exposed terminal problem and explain the difference between the existing MAC protocols for wireless and wired networks, to explain the channel capture effect on MAC and Transport layers, and to describe the concept of proactive and reactive routing protocols. Furthermore, they should be able to describe the concept of geographical and content based routing, and problems of transport layer protocol over multihop wireless networks and present existing solutions. Lastly, but not least, they should be able to describe a simulation scenario and plan the simulation experiments like TCP and UDP protocols over different types of ad hoc routing protocols in a network simulator and measure throughput, packet loss rate performance characteristics. They should also be able to explain the functionality of LTE, WiMax, Zigbee and Bluetooth network architectures. Teaching methods include traditional lectures by academic staff members, guest lectures by industry representatives, lab assignments, and seminar presentations by students. The course takes advantage continual examination with smaller exams given throughout the course.

- Cloud services (7.5 ECTS) – ICT and ICT for Greening

  The learning outcome of this course covers knowledge about scientific foundation in the area of cloud services with proven experience of programming using relevant technologies and devices. Focus areas include cloud services and applications, cloud platforms, and cloud orchestration. The course trains students in their capacity for carrying out teamwork and collaboration with various constellations. Furthermore, the goal is to have students capable of creating, analysing and critically evaluating various technical solutions in terms of the design and implementation of cloud systems and to show insight in research and development by understanding limitations and possibilities. Last, but not least, this course teaches students to plan and use appropriate methods to undertake advanced programming tasks within predetermined parameters and show the ability to identify knowledge gaps and bridging these gaps by gaining new knowledge. Teaching methods include traditional lectures by academic staff members, lab assignments, and seminar presentations by students. The course is finished with a home exam.

- Special Studies in Pervasive and Mobile Computing (Project) (3 ECTS) – ICT and Sustainability

  The learning outcome of this course covers practical aspects of group project where the assignment is a current and interesting problem in the area of distributed computing systems for sustainable development. Students are introduced in Agile development and expected to use the knowledge they have acquired in previous courses and search and study literature to solve the task. So far, students within PERCCOM have been participating in the Green Code Lab Challenge collaborating within teams composed of members both in Luleå, Sweden, and Nancy, France. Teaching methods include traditional lectures by academic staff members, guest lectures by industry representatives, and the above-mentioned project assignment.

- Swedish for Beginners AI:1a - (3 ECTS) – Sustainability (cultural aspects)

  The students start semester 3 with an intensive course in Swedish with the learning outcome to help them in their daily-life in Sweden so that the student can present themselves and their background in Swedish, be able to read and understand simple Swedish texts, and be able to make use of basic knowledge about the structure of the Swedish language. Teaching methods include traditional lectures by academic staff members. In addition, cultural visits are performed during this course. The course is finished with a traditional written exam.

- Seminar: (1.5 ECTS) – ICT and Sustainability

  The learning outcome of this course is the aggregated content of a set of guest lectures by invited academic staff members and senior executives within the IT industry. At the end, students present the status of their own thesis work and submit summaries of the seminars.

After the semester 3 the students are expected in sense of sustainable development to be able

- understand the nature of distributed services and their implementation in modern wireless environments. Students should also see the possibilities of sustainable development in these.
- understand the technologies behind cloud services and possibilities for sustainable development.

Semester 4 - Master's thesis

Although the fourth semester is solely reserved for the Master's thesis, the selected students start their thesis work already during the first semester. As each student has a defined thesis from the first semester on and they will have a possibility to work on their thesis topic on all semesters, the PERCCOM students are expected to show all the expected skills of PERCCOM programme, especially to combine sustainability elements into their thesis topic domains.

1.3.3 Students

PERCCOM programme was actively advertised through EU portals for education and research as well as through research networks and contacts. The programme was a success in many ways from the beginning as shown in Table 1.1. The number of applicants outnumbered the available positions 10–20 times which meant that the programme had the chance of selecting the very best students. One should note here that the EU rules limited the number of selected/supported applicant per country to maximum of 2. This lead to a situation in which some good applicants were put to a reserve list or were offered a position with self-paid option. What is truly exceptions to engineering programme is the high percentage of females as applicants. Although the percentage of female applicants is approximately on the same level with the traditional engineering programmes it is much

Table 1.1   Overview of the student base of PERCCOM programme

Cohort	Applicants Female/Male (Total)	Selected Female/Male (%)	Started Female/Male (%)
2013–2015	78/368 (456)	6/12 (33%)	6/11 (35%)
2014–2016	27/155 (182)	2/12 (20%)	4/11 (26%)
2015–2017	56/235 (291)	8/7 (53%)	7/11 (38%)
2016–2018	47/164 (211)	4/6 (40%)	7/14 (33%)
2017–2019	55/175 (230)	3/7 (30%)	6/13 (31%)

higher in the selected and started students. It seems that the sustainability as a soft engineering approach attracts much more good female applicants than traditional engineering programme.

1.4 Outcomes and Lessons Learned

The outcomes of the PERCCOM programme can be evaluated based on the produced outputs. One of the key indicators of the education programmes is the number of graduating students, their thesis works, their reflection towards the programme as well as their future careers. We analyze here a bit of the thesis works produced in the programme, student careers after the programme as well as their reflections of the programme.

1.4.1 Thesis Works

All five cohorts of students for the PERCCOM programme has been selected and four first cohorts have already graduated. The graduation rate has been rather good. On average only one student per cohort has cancelled the education for different reasons. All the others have managed to graduate in two years like expected. One special feature of the PERCCOM programme was that the thesis topics were given to the students already during the first semester in France, thus the students were able to focus on their thesis research for the full time, although only the last semester was dedicated fully to the thesis works.

The list of all thesis topics is shown in the Appendix A to demonstrate the multitude of effects ICT may have in sustainable development. One can see in the list thesis works with direct greening impacts to the technology, indirect impacts through ICT solutions as well as structural impacts with the technological solutions. Table 1.2 gives some examples of thesis works in each category.

One can see that the work emphasizing the direct sustainability impacts are related to making some technology more efficient in sense of energy usage or carbon emissions. In our examples the efficiency of network routing, mobile network communication and cloud infrastructures are given. Each and every one of these examples benefit of sustainable design. Our examples on indirect sustainability impacts include home environment, transportation and a smart city. First two are heavily linked to the user activities while the third emphasizes more of the technologies in achieving sustainable impacts.

Table 1.2 Examples of thesis topics

Some examples of thesis work with direct impact

- Benchmark of Routing Protocols Regarding Green Considerations
- Ultra-Dense Deployment for Multi Connectivity and Energy Efficiency
- Energy-efficient Cloud Infrastructure for Iot Big Data Processing

The following thesis works represent the examples of indirect effects

- Greenbe – A System to Capture and Visualize Users' Energy-related Activities for Facilitating Greener Energy Behaviour
- Online Transportation Mode Recognition and An Application to Promote Greener Transportation
- Context-driven visualisation of air pollution heat maps with OpenIoT

Finally, structural changes can be seen for example in the following thesis works

- Gamified Participatory Sensing: Impact of Gamification on Public's Motivation in A Lake Monitoring Application
- A Web-based Environmental Toolkit to Support Small and Medium-sized Enterprises in The Implementation of Their Own Environmental Management System
- MEEDS – a Decision Support System for Selecting the Most Useful Developmental Projects in Developing Countries-Case of Ghana

Table 1.3 Statistics of PERCCOM students

- A good balance between companies and academia: 52% of students found a job in companies and 35% found PHD position + 6% with academic positions
- Around 58% of first and second cohort students continued on PhD studies
- The majority of former students stays in Europe (80%)
- North of Europe has been the best place to find jobs in Europe: 44 % of former students are in Sweden, Finland, Netherland, Denmark and Germany.

The examples focusing on systemic effects show behavioural changed in society, enterprises, and project organizations. Common to all these thesis works is that they aim at increasing the knowledge to enable the changes.

1.4.2 Statistics of the Students and their Careers

The PERCCOM programme has been following the careers of the graduated students very closely to learn of the applicability of the education. Appendix B presents the career immediately after the PERCCOM programme and the affiliation of the students. These careers are analysed shortly in the following Table 1.3.

Table 1.4   Results of the student survey

Reason to apply

- Relevance of the programme on ICT and sustainability
- International program within different cultures
- Diversity of the institutions and topics
- Uniqueness and timeliness of the programme

Key outcomes of the programme contents

- Hard skills related to ICT
- Communication (both oral and written) skills
- Adaptation skills
- Teamwork
- Sustainability
- Responsibility

Effects on career

- Good basis for PhD studies
- Good basis to industry work
- Cultural aspects
- Not much

1.4.3 Student Statements Concerning the Programme

Students of the PERCCOM programme were surveyed of their perception to the programme. We wanted to find out (a) what was the most interesting element of the programme that made them apply, (b) how did the contents matched their expectations and (c) how did the programme affected their later career. Survey was answered by 31 students (approximately 35% of all students) and all cohorts were represented in the answers (3 to 9 answers per cohort). The results of this survey are shortly presented in Table 1.4.

1.4.4 Lessons Learned for Sustainability Education

There are several lessons we have learned during the implementation of this programme related to the programme structure and contents, student selection and their backgrounds as well as applicability of sustainable development mind set in various fields. These lessons and guidelines we may give are presented in the Table 1.5.

Table 1.5   Lessons learned from the PERCCOM programme

Programme structure and contents

- Mastering a sustainable (ICT) programme is not feasible for a single partner as the field is so broad. If doing things alone one can, at best, approach some modular structure. By connecting several partners together in education, the programme can better cover the various sustainability aspects.
- High level understanding and vision is more important than the exact set of courses included. Courses as well as technologies will evolve/change while the general mindset will remain. Having a set of partners that complement each other's in sense of sustainability makes a good starting point. At the moment sustainable development is in stage in which earth needs education, innovation and solutions in different domains. This cannot be achieved if the focus is too narrow. One should, however, consider the fact that increased knowledge will eventually change one's perceptions.
- If the sustainable development is linked with ICT then the education is based on ICT (computer science, software engineering etc.) that is complemented with sustainability topic. This means that the teachers in ICT field need to have the ability to integrate the sustainable development to the courses. That will take time. The whole PERCCOM team learned continuously and some perceptions even changed/evolved during this programme.

Student selection and importance of backgrounds

- Sustainable (ICT) programme will attract different types of applicants than the similar kind of a programme without the sustainability aspects. First of all, in ICT field the sustainable development attracts much more female students than traditional ICT programmes. For example, PERCCOM programme selected 30–40% female students while in traditional programmes the percentage is 5–10%. Secondly the background of students varies much more than in traditional programmes (from automation though computer science and engineering to business topics). Although the diversity of students is good, the high vision of the programme should be kept in mind while selecting students. In the end the students should be able to match the learning outcomes of the programme.

Applicability of sustainability in different domains

- Thesis projects have shown us that sustainability can be applied to various domains. Sustainable development is not only the optimization of some ICT technology (greening) but can be the use of ICT in other fields even the human behaviour.
- Increasing the knowledge on sustainable development is the key to better future. Although PERCCOM programme directly affected the knowledge of the students in the programme, it also indirectly affected industries, companies and societies involved in thesis projects.

1.5 Conclusion

As we already highlighted, it is not an easy task to build and run such programs in the absence of standards and clear guidelines. Traditional education standards and curriculum development tools such ACM/IEEE curriculum guidelines for computer science and software engineering education as well as SWEBOK (Software Engineering Body of Knowledge) do not provide any indication on sustainability, green IT or green software engineering. Research and investigations are needed to building a stranded curriculum as universities are running programs and providing feedback to the community. This chapter is a contribution to these efforts.

Sustainability in academic research and education is rather young area as it really started to evolve in the beginning of this millennium and as such the education of sustainable ICT is also evolving. At the starting time of PERCCOM programme, there were no programmes focusing on ICT and sustainability aspects. Although some research groups have been active in publishing articles concerning ICT and sustainable development the research results have not got into actual education. PERCCOM program, as the first of the kind, is by for not perfect. Each hosting partner had in the beginning its own perspective to sustainability and the implementation of sustainability within courses varied. With the first set of students, the PERCCOM consortium has evolved and the PERCCOM program is approaching something that could be multiplied to other locations and other networks. This chapter has presented few lessons learned from the structure and contents of the whole programme but still miss deeper analysis of single courses. This can be done in future publications. What is clear so far is that the highly motivated and talented students and well collaborating universities can create sustainability programme with almost multidisciplinary support. Although we emphasized rather broad vision of sustainable ICT education, in future these programmes will probably focus on a bit more condensed topics, e.g., green networking, sustainable data centres, sustainable software etc. This may take time and few more general programmes like PERCCOM to increase the high-level vision but in the end all disciplines need to adapt to the requirements set by sustainability challenges.

Acknowledgements

The authors would like to thank the support of the European Erasmus Mundus programme PERCCOM (Pervasive Computing and Communications for sustainable development) and the students and staff in this programme and, the Erasmus+ Programme of the European Union.

Appendix A. Thesis Works of the PERCCOM Programme

Thesis Topic	Location
Power Consumption Measurement and Scenario based Energy Model in Wearable Computing Applications	Bremen
Load Balancing in P2p Smartphone based Distribution System	CSIRO
Analysing and Computing the Sustainable Gains of Building Automation	Harz
Development of an Ecology-oriented Software-defined Networking Framework	ITMO
Implementing Green it Approach for Transferring Big Data Over Parallel Data Link	ITMO
A Web-based Environmental Toolkit to Support Small and Medium-sized Enterprises in the Implementation of their Own Environmental Management System	Leeds
Energy Consumption of Applications on Mobile Phones	Leeds
A Bayesian Approach for Forecasting Heat Load in a District Heating System	LTU
Sensor Communication in Smart Cities and Regions: An Efficient IoT-based Remote Health Monitoring System	LTU
Cloudsimdisk: Energy-aware Storage Simulation in Cloudsim	LTU
Sustainable Computer Science Education	LUT
Green Aspects Study in Game Development	LUT
Green Ict Metrics and Biomimicry	UL
Benchmark of Routing Protocols Regarding Green Considerations	UL
Green Service Level Agreement Under Sustainability Lens in it Industry	UL
Modelling Energy Consumption of a Switch using Fuzzy-rule Classifier	UL

(*Continued*)

Thesis Topic	Location
Analysing the Power Consumption Behaviour of Ethernet Switch using Design of Experiment	UL
Optimizing Last Mile Delivery using Public Transport with Multi- Agent based Control	Bremen
Nested Rollout Policy Adaptation For Optimizing Vehicle Selection in Complex Vrps	Bremen
Mobile-GSN (Mgsn) for Enhanced Sensor Management	CSIRO
Reasoning Over Knowledge-based Generation of Situations in Context Spaces to Reduce Food Waste	CSIRO
RCOS: Real time Context Sharing Across a Fleet of Smart Mobile Devices	CSIRO
Front-end Development for Building Automation Systems using JavaScript Frameworks	HARZ
Specification of a Smart Meter/Actuator Description Mechanism & Development of a Mobile Application	HARZ
Embedding Sustainability into the New Computer Science Curriculum for English Schools	Leeds
A Belief Rule-based Environmental Responsibility Assessment System for Small and Medium-sized Enterprises	Leeds
The Viability of a Tool for Fetal Health Monitoring	LTU
Performance Analysis of IP Based WSNs in Real time Systems	LTU
Early Investigation towards Defining and Measuring Sustainability as a Quality Attribute in Software Systems	LUT
Cyber Foraging for Green Computing, Improving Performance and Prolonging Battery Life of Mobile Devices	LUT
Developing Strategies to Mitigate the Energy Consumed by Network Infrastructures	UL
Using Ict Energy Consumption for Monitoring ICT usage in an Enterprise	UL
Distributed Context Acquisition and Reasoning in the Internet of Things for Indoor air Quality Monitoring	CSIRO
Opportunistic Collection of Sensor Data in IoT Applications	CSIRO

Thesis Topic	Location
Remote Control Based Home Automation Usability Evaluation	Harz
Improving Energy Efficiency of Residential Building Automation System Considering user Contexts Enriched by Smartphones - German use Case -	Harz
Pervasive Computing for Decision Support Systems in the Context of Green ICT	ITMO
A Smart Waste Management System using IoT and Blockchain Technology	ITMO
Sustainability of Greenmed: a Physical Activity Monitoring Application	Leeds
Performance Evaluation of Scalable and Distributed IoT Platforms for Smart Regions	LTU
Ultra Dense Deployment for Multi Connectivity and Energy Efficiency	LTU
Online Transportation Mode Recognition and an Application to Promote Greener Transportation	LTU
An Intelligent Flood Risk Assessment System using Belief Rule Base	LTU
Engineering and Incorporating Sustainability into Software Development: a Design Pattern Approach	LUT
Greenbe – a System to Capture and Visualize users' Energy-related Activities for Facilitating Greener Energy Behaviour	LUT
Gamified Participatory Sensing: Impact of Gamification on Public's Motivation in a Lake Monitoring Application	LUT
Assessing the Benefit of Deploying EE on Commercial Grade Network Switches	UL
Optimizing the Energy Consumption of Core Networks with Software Defined Networks Perspective	UL
Analysing API Calls to Reduce Energy Consumption of Apps in Idle States	UL
Capacity Management in Hyper-scale Datacenters using Expert System/Machine Learning	LTU

(*Continued*)

Thesis Topic	Location
Machine Learning Assisted System for the Resource-constrained Atrial Fibrillation Detection from Short Single-lead Ecg Signals	LTU
Blixt^{TM}: an Available Bandwidth Measurements' Approach for High-speed Mobile Networks	LTU
Defining Quality of Experience in Industrial Internet of Things	LTU
A Belief Rule Based Flood Risk Assessment Expert System using Real time Sensor Data Streaming	Chittagong
Using Gestures to Interact with Home Automation Systems	Harz
Comparing Javascript Frameworks and Dart for Front-end Development in Building Automation	Harz
Architecting and Designing Sustainable Smart City Services in a Living Lab Environment	LUT
A Framework and a Web Application for Self- Assessment of Sustainable Green ICT Practices in SMEs	LUT
Quality of Experience Provisioning in Mobile Cloud Computing	LUT
MEEDS -a Decision Support System for Selecting the Most useful Developmental Projects in Developing Countries- Case of Ghana	LUT
Successful Patterns in Corporate Social Responsibility in Information Technologies	CSULB
Green Software Defined Distributed Data Center Model	LBU
Energy-efficient Cloud Infrastructure for IoT Big Data Processing	LBU
Budget of Iot Low Power Wide Area Network Architectures	UL
Green Grid Metrics in Network Information System for Monitoring Ict Sustainability in Data Centers	UL
New Multimedia Services in SDN Considering Energy Consumption of the Network	UL
Data Center Performance Comparison Framework Based on Biomimicry	UL
Performance Evaluation of IoT Platforms in Green Ict Applications	CSIRO

Thesis Topic	Location
Generating O-mio-df Wrappers for Connected Smart Objects in Green ICT Applications	CSIRO
Performance Evaluation of Messaging System for IoT Enabled Waste Management System	ITMO
An Ontology Driven Model for Environmental Legislations	LBU
Matching Data Centre Energy Consumption with Renewable Energy Mix Generation	LBU
5g Multi−connectivity, Ue Performance for Dense Urban City Scenarios	LTU
Green Ai with Binarized Neural Network Architectures	LTU
Big Data Orchestration Framework for Mobile Edge Computing	LTU
5g & Brb based Deep Learning Approach with Application in Sensor Data Streams	Chittagong
Big Data Meet Green Challenges: Greening Big Data	UL
The Role of Data Center in Smart Cities	ENEA
NFV Deployment Scenario Dimensioning	UL
Context- and Situation-prediction for Outdoor Air Quality Monitoring	CSIRO
Context-driven Visualisation of Air Pollution Heat Maps with OpenIoT	CSIRO
Proactive Adaptation of Behaviour for Smart Connected Objects in Biotope Ecosystem	ITMO
Analysis and Improvisation of Data Quality in Citizen Science	LUT
Affecting the User's Sustainability Behaviour	LUT
Modelling and Analysing Sustainability Effects of Various Modes of Transportation in Finland	LUT
Optimization of Wired/Wireless Network Gateways Placement in Iot	UL
Analysing & Computing the Gain (Level of Sustainability) of Stand-by Cut-offs	Harz
Improving the Effectiveness of Building Automation by Adaption to the users Context	Harz
Modelling of Secure Communication System for IoT Enabled Waste Management System	ITMO

Appendix B. Careers by Graduated PERCCOM Students (First 3 Cohorts)

Where	Organization	Position
Saga, Japan	Saga University	PhD Student (graduated 2018)
Odense, Denmark	University of Southern Denmark	PhD Student (graduated 2018)
Cluj, Romania	Garmin	Developer
Geneve, Switzerland	University Of Geneva	PhD Student
Jakarta, Indonesia	Telkom Telstra, a joint venture of Telco company	Senior Officer
Athens, Greece/ Netherlands	Athens University of Economics and Business, Delft Technological University	PhD Student
Dhaka, Bangladesh	University of Asia Pacific, Dhaka	Lecturer
Paris, France	Brinks France	Network system administrator
Helsinki, Finland	Futudent (Novocam Medical Innovations Oy /Ltd)	Product Support & Customer Care
Bologna, Italy	University of Bologna, University of Turin	PhD Student
Schipol, Amsterdam, Netherlands	Cargill	IT Management trainee
Castellon, Spain	Universitat Jaume I, GEO-C	PhD Student
Lappeenranta, Finland	Lappeenranta University of Technology	PhD Student
Nancy, France	University of Lorraine, CRAN	PhD Student (graduated 2018)
Sophia Antipolis, France	Sophia Antipolis, I3S	PhD Student

Where	Organization	Position
Jabodetabek, Indonesia	Niometrics	DevOps
Berlin, Germany	simpleTechs GmbH	Project Manager
Bremen, Germany	University of Bremen	Graduate student researcher
Malmo, Sweden	EON	Graduate IT trainee
Bilbao, Spain	University of Deusto	PHD fellow
Helsinki, Finland	Aalto University	PHD student
Eindhoven, Netherlands	TOPIC Embedded Systems	Software engineer
Finland	Löorn Solutions Oy	CEO
Bangladesh	East West University,	Lecturer
Luxembourg	University of Luxembourg	PHD Student
Lulea, Sweden	Lulea University of Technology	PHD Student
Lappeenranta, Finland	Lappeenranta University of Technology	PHD student
Mauritius	Accenture Technology	Software Developer
Europe	–	Looking for Network Engineer position
Barcelona, Spain	Universitat Politècnica de Catalunya	PHD student
Gothenburg, Sweden	Volvo Trucks	Embedded Software Engineer
Stockholm, Sweden	Netlight Consulting	Analyst
Stockholm, Sweden	Ericsson	Linux Developer
Gothenburg, Sweden	Chalmers	Ph.D student
Munich, Germany	Adnymics	Platform Engineer
Open Lab Summer Intern, Switzerland	CERN	Intern (Jul.–Sept 2017)
Aalborg, Denmark	Erasmus It4bi-dc	Ph.D student
Lappeenranta, Finland	Visma	Mobile Developer

(Continued)

Where	Organization	Position
Helsinki, Finland	Tieto	Java Developer
Stockholm, Sweden	Ericsson	Researcher on Cloud Service Assurance
Curitiba, Brazil	Verti Tecnologia	Full Stack developer and Embedded developer
Australia	–	Looking for a PHD in Australia
Brugge, Belgium	–	Looking for work
Dusseldorf, Germany	JobTender24	Web Developer
Luxemburg	RTL Group	Information Security Trainee
Paris, France	Botify	Javascript Engineer
Stockholm, Sweden	Ericsson	Software developer
Atlanta, USA	Stratocore	Consultant

References

Brundtland, G., Khalid, M., Agnelli, S., Al-Athel, S., Chidzero, B., Fadika, L., Hauff, V., Lang, I., Shijun, M., Morino de Botero, M., Singh, M., Okita, S., Others, A., 1987. Our Common Future ('Brundtland report'), Oxford Paperback Reference. Oxford University Press, USA.

Cai, Y., 2010. Integrating sustainability into undergraduate computing education, in: Proceedings of the 41st ACM Technical Symposium on Computer Science Education. ACM, pp. 524–528.

Computer Science Curricula 2013 Curriculum Guidelines for Undergraduate Degree Programs in Computer Science 2013 (Final Report), 2013. The Joint Task Force on Computing Curricula Association for Computing Machinery IEEE-Computer Society.

Down, L., 2006. Addressing the challenges of mainstreaming education for sustainable development in higher education. Int J Sustain. High. Educ. 7, 390–399.

Falkenberg, T., Babiuk, G., 2014. The status of education for sustainability in initial teacher education programmes: A Canadian case study. Int J Sustain. High. Educ. 15, 418–430.

Gibson, M.L., others, 2017. Mind the chasm: A UK fisheye lens view of sustainable software engineering, in: 6th Intl. Workshop on Requirements Engineering for Sustainable Systems. pp. 15–24.

Hilty, L.M., Aebischer, B., 2014. ICT Innovations for Sustainability. Springer Publishing Company, Incorporated.

Horhota, M., others, 2014. Identifying behavioral barriers to campus sustainability: A multi-method approach. Int J Sustain. High. Educ. 15, 343–358.

Klimova, A., Rondeau, E., Andersson, K., Porras, J., Rybin, A., Zaslavsky, A., 2016. An international Master's program in green ICT as a contribution to sustainable development. J. Clean. Prod. 135, 223–239. https://doi.org/10.1016/j.jclepro.2016.06.032

Mann, S., 2016. Computing Education for Sustainability: What Gives Me Hope? interactions 23, 44–47.

Mann, S., 2011. The Green Graduate: Educating Every Student as a Sustainable Practitioner. ERIC.

Mann, S., Muller, L., Davis, J., Roda, C., Young, A., 2010. Computing and sustainability: evaluating resources for educators. ACM SIGCSE Bull. 41, 144–155.

Mann, S., Smith, L., Muller, L., 2008. Computing education for sustainability. ACM SIGCSE Bull. 40, 183–193.

Özkan, B., Mishra, A., 2015. A curriculum on sustainable information communication technology. Probl. Sustain. Dev. 10.

Penzenstadler, B., Betz, S., Venters, C.C., Chitchyan, R., Porras, J., Seyff, N., Duboc, L., Becker, C., 2018. Everything is INTERRELATED: teaching software engineering for sustainability, in: Proceedings of the 40th International Conference on Software Engineering: Software Engineering Education and Training. ACM, pp. 153–162.

Penzenstadler, B., Fleischmann, A., 2011. Teach sustainability in software engineering?, in: Intl. Conf. on Software Engineering Education and Training. IEEE.

Rockström, J., Steffen, W., Noone, K., Persson, Å., Chapin III, F.S., Lambin, E.F., Lenton, T.M., Scheffer, M., Folke, C., Schellnhuber, H.J., Nykvist, B., de Wit, C.A., Hughes, T., van der Leeuw, S., Rodhe, H., Sörlin, S., Snyder, P.K., Costanza, R., Svedin, U., Falkenmark, M., Karlberg, L., Corell, R.W., Fabry, V.J., Hansen, J., Walker, B., Liverman, D., Richardson, K., Crutzen, P., Foley, J.A., 2009. A safe operating space for humanity. Nature 461, 472.

Rusinko, C.A., 2010. Integrating sustainability in higher education: a generic matrix. Int. J. Sustain. High. Educ. 11, 250–259.

Software Engineering 2014 Curriculum Guidelines for Undergraduate Degree Programs in Software Engineering, 2014. Joint Task Force on Computing Curricula IEEE Computer Society Association for Computing Machinery.

Tainter, J., 2006. Social complexity and sustainability. J. Ecol. Complex. 3, 91–103. https://doi.org/10.1016/j.ecocom.2005.07.004

Tangwanichagapong, S., others, 2017. Greening of a campus through waste management initiatives: Experience from a higher education institution in Thailand. Int J Sustain. High. Educ. 18, 203–217.

Weenen, H. van, 2000. Towards a vision of a sustainable university. Int J Sustain. High. Educ. 1, 20–34.

2

Review of Decision Support Methods in Green and Sustainable Supply Chains

Davidson de Almeida Santos[1*], Osvaldo Luiz Gonçalves Quelhas[1], Carlos Francisco Simões Gomes[1], Rodrigo Goyannes Gusmão Caiado[1] and Sheila da Silva Carvalho Santos[2]

[1]Fluminense Federal University, Brazil
[2]Celso Suckow da Fonseca Federal Center for Technological Education, Brazil
E-mail: davidsonadm@yahoo.com.br; osvaldoquelhas@id.uff.br; cfsg1@bol.com.br; rodrigocaiado@id.uff.br; sheila.santos@cefet-rj.br
*Corresponding Author

We present a systematic literature review of the application of decision support techniques, especially the multi-criteria decision making (MCDM) methods in green and sustainable supply chain management (G/S – SCM) in Scopus database. After defining some filtering criteria, we obtained 65 articles that were analysed based on the following items: quantitative of publications per year; number of articles per newspaper; number of articles per country; number of articles per Educational Institution; number of authors per article; number of authors with greater representativeness; number of citations per article; quantitative methods used in the articles; number of types of decision support methods and the purpose of applying the MCDM methods. The main findings of the paper were: the formation of an overview of the applications of decision support methods in green and sustainable supply chains; the survey of supplier selection/evaluation criteria and performance measures aimed at green and sustainable chains. The results obtained from the analyses will allow for further studies, applications of other decision support methods in green and sustainable supply chains; use of supplier selection criteria for green and sustainable chains; selection of performance measures for the evaluation of green and sustainable chains; and use of the MCDM to meet other environmental or sustainability purposes.

2.1 Introduction

Supply chain sustainability has been highlighted as a precursor to sustainable manufacturing and sustainable development. Manufacturing companies are considered sustainable, as long as their supply chains are also sustainable. Supply chain sustainability in manufacturing companies usually involves a wide range of stakeholders, both within and outside the organization (Su et al., 2016).

Due to increased knowledge of customer needs and the ecological pressures of markets and other stakeholders (e.g., company shareholders, suppliers), organizations understood the importance of greening and sustainability in their supply chain through of supplier selection (Luthra et al., 2017). These challenges are becoming more visible as there is growth in industry sectors where competing firms have 222 similar access to sources of raw materials, technology, and similar suppliers. Continued depletion of natural resources, destruction of the quality of the natural environment, raising consumer and other stakeholder awareness, and increasingly stringent emission regulations imposed by the authorities are some of the factors that compel policy makers to seek inclusive social solutions and economic and environmental growth, a three-dimensional view of sustainability known as the triple bottom line (TBL) (Kruger et al., 2018).

Companies are continually being challenged to strategize and develop new capabilities such as clean manufacturing, green product design and waste disposal, and to achieve the kind of development that can be sustained for future generations as well. Companies responded by adopting more integrated and systematic methods to improve sustainability performance, which laid the foundation for new business models (Su et al., 2016). Amidst these new challenges comes the concept of green and sustainable supply chain management (G/S-SCM). All industries, regardless of their size, seek improvements in their processes: raw material purchase, manufacturing, allocation, transport efficiency, reduction of storage time, import and disposal of products, aiming to meet environmental objectives and reduce costs in the manufacturing process (Kannan et al., 2015).

The increasing importance of G/S-SCM is driven mainly by the increasing deterioration of the environment, for example, reduction of raw material resources, overflow of waste sites and increase of pollution level. G/S-SCM originates from the literature on environmental management and supply chain management. The inclusion of the "green" component in supply chain management signals the influence and relationships between supply chain management and the environment (Chithambaranathan et al., 2015). In this sense, this article aims to carry out a systematic review of the applications of multi-criteria decision-making – MCDM methods in green and sustainable supply chains in order to answer the following question: what are the main criteria for supplier selection/evaluation and performance measures used for G/S-SCM?

The chapter will be divided as follows: the second section presents the methodology used to select the article base; the third refers to the analysis of descriptive results and the fourth the thematic synthesis generated from the process of systematic literature review. The fifth section corresponds to the conclusions and suggestions for future studies.

2.2 Methodology

The research is classified as exploratory, because it aims at a greater under-standing about a given problem. For this a bibliographical survey is carried out. In this case, the research involves both qualitative approach, admitting that the subjectivity of the authors is present in the several stages of the liter-ature review process, and quantitative approach through Bibliometrics – use of statistical techniques for dissemination of scientific knowledge (Araújo, 2006) – for the identification of trends and relevant MCDM methods. In addi-tion, it is possible to classify the research as descriptive, since the structure that allows the selection of the articles is presented.

Concerning the systematic literature review (SLR), it was based on Caiado et al. (2018) method, consisted of four steps: (1) formulating ques-tion(s) for the research; (2) selection and evaluation of studies; (3) analysis of the content of selected articles; and, (4) the description of results. The search is limited to a set of keywords ('Multiple Attribute', 'Multi-Attribute', 'Decision Make', 'Multiple Criteria', 'Multi-Criteria', 'Multi Criteria', 'Decision Support', 'Multiple Objective', 'Multi-Objective', 'Multiple objec-tive', 'Multiple Objective Decision Aiding', 'Multiple Criteria Decision Aid') in the Scopus (scopus.com) database. The conducted research had combined these keywords into title, abstract or keywords, limited to articles, in English, published in peer-reviewed journals from 2014 up to July 2018, from Busi-ness, Management and Accounting or Decision Sciences areas. The SLR process is indicated in Figure 2.1.

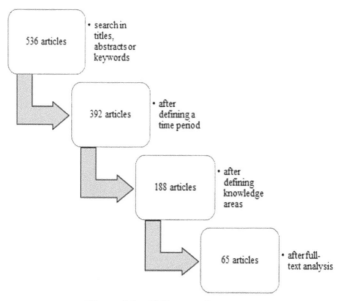

Figure 2.1   SLR research process.

Based on the full text analysis, a total of 65 articles complied with the selection criteria. Hence these were all the articles that, to a certain extent, referred to multi-criteria decision making methods in green and sustainable supply chains. In the next stage, researchers discussed and created tables using Microsoft Excel.

The description of results consisted of two analyses. Firstly, there was a descriptive analysis, pointing out the quantitative of: publications per year; authors with greater representativeness; methods used in the articles; types of multi-criteria decision making methods and also for the purpose of applying the MCDM method. Then, there was a thematic synthesis analysis, in which individual articles were categorized and organized by criteria for vendor selection/evaluation and G/S-SCM performance measures.

2.3 Descriptive Results

Figure 2.2 shows the number of publications per year and portray that although the year 2017 has suffered a decrease compared to the previous

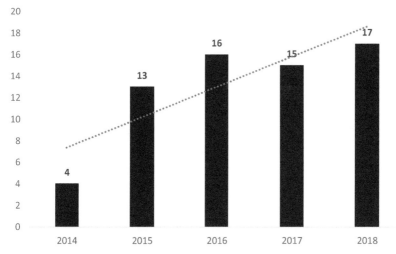

Figure 2.2 Number of publications per year.

year (2016); there is a growing trend in the number of publications over the years.

Then, Figure 2.3 presents the number of articles per journal, which represents the most relevant periodicals in the dissemination of studies related to applications of MCDM in green and sustainable supply chains.

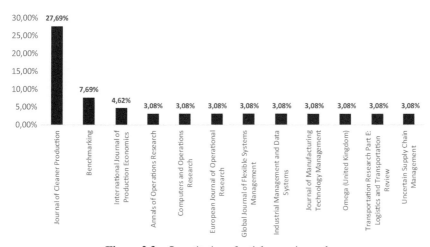

Figure 2.3 Quantitative of articles per journal.

According to Figure 2.3, the Journal of Cleaner Production presented the highest percentage of article publications (18 articles) among the 33 journals identified in the survey. It should be noted that the first 12 journals represented in the graph correspond to 67.69% of the number of articles used to form the analysis base.

The number of authors with greater representativeness (Figure 2.4) presents the authors who developed the highest index of studies related to the applications of multi-criteria decision making methods in green and sustainable supply chains. Noteworthy are the authors: Kannan Govindan who participated in 10 articles (as author or co-author); Joseph Sarkis with 4 articles (as author or co-author) and the other authors mentioned in Figure 2.4 with 3 articles each.

In addition, the percentage distribution of articles by country identified the countries that are developing more studies related to applications of decision support methods in green and sustainable supply chains.

Figure 2.5 shows the percentage distribution of articles by country, showing that the concentration of scientific production is between India and Iran representing 55.38% in total.

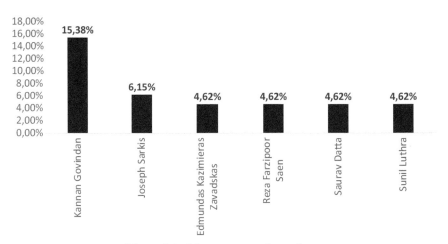

Figure 2.4   Most representative authors.

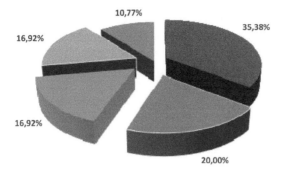

10,77%

35,38%

16,92%

16,92%

20,00%

■ India   ▪ Iran   ▨ China   ▪ Denmark   ■ United States

Figure 2.5   Quantitative of articles by country.

Moreover, we also identified the main Educational Institutions (EIs) that develop studies related to decision support methods applied to G/S-SCM. Figure 2.6 presents the quantitative of articles by EIs, pointing out three institutions that are prominent in the development of studies, namely: University of Southern Denmark (Denmark), National Institute of Technology (India) and Islamic Azad University (Iran). It should be noted that the eight institutions in the chart below correspond to 58.46% of the articles in the base analysis for the survey.

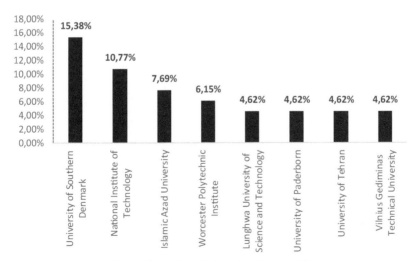

Figure 2.6   Quantitative of articles by EI.

The number of authors per article identifies the number of authors involved in the elaboration of articles. In order to identify the number of authors involved in the elaboration of the articles, Figure 2.7 shows that most articles (75.38%) were developed by three or more authors.

Besides, we also analyse the number of methods used in the articles (Figure 2.8) in order to check the number of MCDM methods used per article.

Figure 2.8 shows that 50.77% of the articles analysed were structured using a MCDM method, 35.38% – two MCDM methods, 9.23% – three MCDM methods and 4.62% – four MCDM methods. In this case there is an opportunity, as yet little explored, for the application of 3 or more multi-criteria decision making methods in green and sustainable supply chains.

The number of MCDM methods shows which methods were used in this article portfolio for the present study (65 articles). Figure 2.9 shows the prominence of the Fuzzy method (55.38% – 36 articles) among the others. The second most used method is TOPSIS (Technique for Order of Preference by Similarity to Ideal Solution) (23.08% – 15 articles), followed by AHP (Analytic Hierarchy Process) (21.54% – 14 articles).

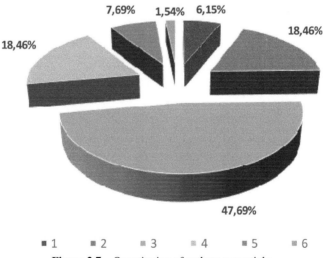

Figure 2.7   Quantitative of authors per article.

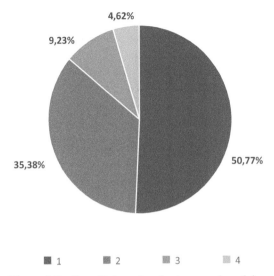

Figure 2.8 Quantitative of methods serves in articles.

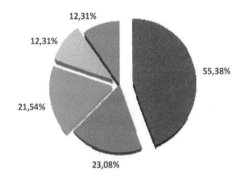

Figure 2.9 Quantitative methods of decision support.

The proposed analysis allows the identification of the opportunity to apply other MCDM methods such as ELECTRE (ELimination Et Choix Traduisant la REalité) and PROMETHEE (Preference Ranking Organization Method for Enrichment Evaluations) in green and sustainable supply chains.

The purpose of applying the MCDM method indicates the motivation that generated the need to use decision aid methods in green and sustainable supply chains. Figure 2.10 shows that 58.46% (38 articles) presented applications of multi-criteria decision-making methods focused on selection and evaluation of supplier performance. Another important point to be highlighted with this analysis refers to the opportunity of applications of MCDM methods with the objective of serving other purposes. It is also observed a high concentration (15.38%) of studies focused on G/S-SCM performance measurements.

2.4 Thematic Synthesis

The analysis developed in the previous section allowed the construction of two tables that concentrate the main supplier selection/evaluation criteria for a green and sustainable supply chains and the main performance measures for G/S-SCM. It should be noted that this survey is directly related to the percentage distribution present in Figure 2.10. The information presented in Tables 2.1 and 2.2 may allow the structuring of a number of other studies related to applications of MCDM methods in G/S-SCM.

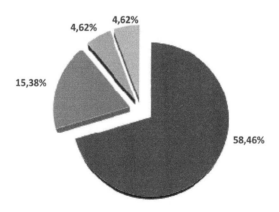

Figure 2.10   Purpose of the application of the MCDM method.

Table 2.1 Criteria for supplier selection/evaluation

Main Criterion	Sub-Criterion	Authors
Quality	Issues related to quality as a system and quality assurance.	Kannan et al. (2015) Wu and Barnes, (2016) Banaeian et al. (2018) Govindan et al. (2017)
Economic	Aspects related to purchase price, transportation cost, profit and financial capacity.	Kannan et al. (2015) Luthra et al. (2017) Wu and Barnes (2016) Banaeian et al. (2018) Govindan et al. (2017) Tavana et al. (2017)
Supplier/ Delivery Capacity	Supply capacity, supplier research and development	Kannan et al. (2015) Luthra et al. (2017) Govindan et al. (2017)
Environment	It involves the following points: certification of environmental protection system; environmental efficiency; ecodesign requirements for energy-using products; environmental protection policies/plans; environmental management systems; environmental costs; environmental competencies.	Kannan et al. (2015) Luthra et al. (2017) Wu and Barnes (2016) Rostamzadeh et al. (2015)
Corporate Social Responsibility (CSR)	The rights and interests of the stakeholders (including the employee).	Kannan et al. (2015) Luthra et al. (2017) Govindan et al. (2017) Tavana et al. (2017)
Pollution control	Air emissions; residual waters; pollution control initiatives; capacity to reduce pollution; prevention and management of waste and pollution; quantity control and treatment of solid waste; quantity control and treatment of hazardous emissions, such as SO_2, NH_3, CO, and HCl; quantity control and wastewater treatment; evaluation if the supplier has environment-related certificates, such as ISO14000;	Kannan et al. (2015) Luthra et al. (2017) Wu and Barnes (2016) Rostamzadeh et al. (2015) Govindan et al. (2017) Tavana et al. (2017)
Green Product	Recycle; green packaging; costs of disposal of components; green certifications; green production; reuse; re-manufacture; disposition.	Kannan et al. (2015) Luthra et al. (2017)

(Continued)

Table 2.1   Continued

Main Criterion	Sub-Criterion	Authors
Green image	Materials used in supplied components that reduce the impact on natural resources; ability to change process and product to reduce impact on natural resources; proportion of green customers to total customers; market share of green customers; stakeholder relationship; encoding and recording of green materials.	Kannan et al. (2015) Luthra et al. (2017)
Green Innovation	Green technological capabilities; green design; green process/production planning; recycling of product design; renewable product design; green research and development project; product redesign; infrastructure for research and development work and capacity to develop new projects and speed of development.	Kannan et al. (2015) Luthra et al. (2017) Govindan et al. (2017) Tavana et al. (2017)

Table 2.2   G/S-SCM performance measures

Performance Measure	Description	Author
Profitability	The company's ability to generate more revenue than net investment. In this case the following criteria could be considered: price strategy; use of assets;	Qin et al., (2017); Wu et al., (2016)
Market share	Proportion of sales controlled by the company in the whole market in which the company operates	Qin et al., (2017); Wu et al., (2016)
Social Performance	Measure how well the company's social goals are successfully put into practice. The following criteria could be inserted: employee practices; reduced community impact; health and safety; laws and regulations; sustainable packaging and improved relationships with community stakeholders and community activists.	Qin et al., (2017); Chithambaranathan et al., (2015); Wu et al., (2016)
Organizational Commitment	It is the psychological attachment of the individual to the organization, predicts work variables such as turnover, organizational citizenship behavior and work performance.	Chithambaranathan et al., (2015)
Green supply chain process	Activities that associate: product design, selection and supply of materials, manufacturing processes, delivery of the product to consumers and end of life of the product after its useful life (sustainability–recycling–reverse logistics).	Chithambaranathan et al., (2015); Wu et al., (2016)

Table 2.2 Continued

Performance Measure	Description	Author
Eco Design	It is an approach to designing a product with special consideration for the environmental impacts of the product throughout its life cycle (Lifecycle Management).	Chithambaranathan et al., (2015); Wu et al., (2016)
Sustainable performance	It is the philosophy of environmental and social sustainability in products as well as production processes. It develops solutions that meet the highest environmental standards.	Chithambaranathan et al., (2015); Wu et al., (2016)
Supplier profile	Refers to the following items: technological maturity; geographical proximity; life cycle cost management; volume of CO_2 related to the request and strategic importance.	Theißen and Spinler, (2014)
CO_2 management skills	Spectrum of CO_2 reduction practices; accounting and reporting of CO_2; definition of CO_2 reduction plans and targets and commitments to external environmental management activities.	Theißen and Spinler, (2014)
Organizational Factors	Proximity of relationship; response to change; management of CO_2 risk assessment; joint investments in CO_2 reduction practices; congruences of CO_2 strategies.	Theißen and Spinler, (2014)
Risk factors	Loss of CO_2 savings from supplier; delays in conducting CO_2 reduction practices and transfer of CO_2 management knowledge to rivals.	Theißen and Spinler, (2014)
Stakeholder	Influence of stakeholders on the GCSV.	(Wu et al., 2016)
Resilience	Corresponds to flexible and clean technology.	(Wu et al., 2016)

2.5 Proposal for a Sustainable Performance Measurement System for Supplier Selection

From Tables 2.1 and 2.2, it was possible to generate a sustainable framework integrated with a systemic view of the selection of the key suppliers in green and sustainable supply chain management considering all levels of classes of suppliers. This framework will be adherent to the three dimensions of sustainability simultaneously.

Hence, we will evaluate the supplier by analysing the efficiency of its processes, analysis of the life cycle of its products and the social responsibility of its practices and projects. In addition, this framework (Figure 2.11) will be a new proposal for a supplier performance evaluation system and will have

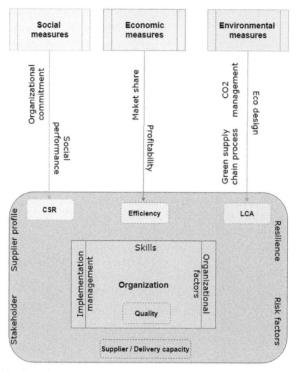

Figure 2.11   Sustainable performance measurement system for supplier selection.

a systemic view, as it will consider the inherent risk types, resilience, supplier profile and stakeholder engagement for selecting the most appropriate supplier of G/S-SCM.

2.6 Conclusions and Suggestions for Further Research

The research identified that there is a year-on-year increase in publications regarding the applications of MCDM methods in green and sustainable supply chains, thus demonstrating a growing interest in the subject. Another important point was the recognition of Kannan Govindan as the main author, with 10 articles (as author or co-author). In addition, the aforementioned author is present in the first three most cited articles and in a systematic review constructed between 2014–2018.

The research also pointed to the possibility of applying three or more MCDM methods and the opportunity to apply new MCDM methods to meet other purposes of green and sustainable supply chains. The article allowed the construction of a reference table for the structuring of new studies aimed at supplier selection/evaluation and performance evaluation of G/S-SCM (Tables 2.1 and 2.2).

The suggestion for future studies is concentrated on the following points: use of three or more MCDM methods in G/S-SCM; application of multi-criteria decision making methods for other purposes (e.g., selection of technology for green and sustainable supply chains) and use of reference tables for the development of new studies.

References

Araújo, C. A. (2006). Bibliometria: evolução histórica e questões atuais. Em questão, *12*(1), 11–32.

Banaeian, N., Mobli, H., Fahimnia, B., Nielsen, I. E., and Omid, M. (2018). Green supplier selection using fuzzy group decision making methods: A case study from the agri-food industry. *Computers & Operations Research*, *89*, 337–347.

Caiado, R. G. G., Leal Filho, W., Quelhas, O. L. G., de Mattos Nascimento, D. L., and Ávila, L. V. (2018). A literature-based review on potentials and constraints in the implementation of the sustainable development goals. *Journal of Cleaner Production*.

Chithambaranathan, P., Subramanian, N., Gunasekaran, A., and Palaniappan, P. K. (2015). Service supply chain environmental performance evaluation using grey based hybrid MCDM approach. *International Journal of Production Economics*, *166*, 163–176.

Govindan, K., Kadziński, M., and Sivakumar, R. (2017). Application of a novel PROMETHEE-based method for construction of a group compromise ranking to prioritization of green suppliers in food supply chain. *Omega*, *71*, 129–145.

Govindan, K., Rajendran, S., Sarkis, J., and Murugesan, P. (2015). Multi criteria decision making approaches for green supplier evaluation and selection: a literature review. *Journal of Cleaner Production*, *98*, 66–83.

Kannan, D., Govindan, K., and Rajendran, S. (2015). Fuzzy Axiomatic Design approach based green supplier selection: a case study from Singapore. *Journal of Cleaner Production*, *96*, 194–208.

Kruger, C., Caiado, R. G. G., França, S. L. B., and Quelhas, O. L. G. (2018). A holistic model integrating value co-creation methodologies towards the sustainable development. *Journal of Cleaner Production*, *191*, 400–416.

Luthra, S., Govindan, K., Kannan, D., Mangla, S. K., and Garg, C. P. (2017). An integrated framework for sustainable supplier selection and evaluation in supply chains. *Journal of Cleaner Production*, *140*, 1686–1698.

Qin, J., Liu, X., and Pedrycz, W. (2017). An extended TODIM multi-criteria group decision making method for green supplier selection in interval type-2 fuzzy environment. *European Journal of Operational Research*, *258*(2), 626–638.

Rostamzadeh, R., Govindan, K., Esmaeili, A., and Sabaghi, M. (2015). Application of fuzzy VIKOR for evaluation of green supply chain management practices. *Ecological Indicators*, *49*, 188–203.

Su, C. M., Horng, D. J., Tseng, M. L., Chiu, A. S., Wu, K. J., and Chen, H. P. (2016). Improving sustainable supply chain management using a novel hierarchical grey-DEMATEL approach. *Journal of Cleaner Production*, *134*, 469–481.

Tavana, M., Yazdani, M., and Di Caprio, D. (2017). An application of an integrated ANP–QFD framework for sustainable supplier selection. *International Journal of Logistics Research and Applications*, *20*(3), 254–275.

Theißen, S., and Spinler, S. (2014). Strategic analysis of manufacturer-supplier partnerships: An ANP model for collaborative CO_2 reduction management. *European Journal of Operational Research*, *233*(2), 383–397.

Wu, C., and Barnes, D. (2016). An integrated model for green partner selection and supply chain construction. *Journal of Cleaner Production*, *112*, 2114–2132.

Wu, K. J., Liao, C. J., Tseng, M., and Chiu, K. K. S. (2016). Multi-attribute approach to sustainable supply chain management under uncertainty. *Industrial Management & Data Systems*, *116*(4), 777–800.

3

Analyzing the Drivers of Engineering Education for Sustainability using MCDM Approach

S. Vinodh[*] and Rohit Agrawal

Department of Production Engineering, National Institute of Technology, Tiruchirappalli-620015, Tamil Nadu, India
E-mail: vinodh_sekar82@yahoo.com; mailerrohit@gmail.com
*Corresponding Author

Sustainable concepts are essential in engineering education to enhance the effectiveness of Teaching-Learning process. In this context, this chapter presents an attempt of identifying drivers of engineering education from sustainability viewpoint and subsequent analysis using a Multi Criteria Decision Making (MCDM) tool. Analytical Hierarchy Process (AHP) is used as solution methodology. 18 drivers are being identified from literature and are prioritized based on expert opinion. The key drivers are identified based on priority order generated from AHP. The inferences based on the study are highlighted.

3.1 Introduction

Sustainability has become an important aspect for product development. According to Glavic (2006) the industries are moving towards pollution prevention rather than pollution control. Deployment of sustainability principles in product design stage is very challenging because of less conceptual knowledge towards sustainability and complexity in the methodology (Hallstedt, 2017). A review showed that educational experts have more concern towards integrating sustainability education in present engineering curriculum so that engineering student can better understand the importance of sustainability and then they can design product keeping sustainability concept in mind (Esparragoza et al., 2018). This situation necessitates understanding the drivers of engineering education for sustainability. In this context, the analysis of drivers of engineering education from sustainability dimension is gaining importance. This chapter presents the analysis of such drivers using Multi Criteria Decision Making (MCDM) approach. Analytical Hierarchy Process (AHP) is used for analysis of drivers. The analysis revealed the top drivers to be focused upon. The unique aspect of the study is that it presents identification and systematic analysis of drivers of engineering education from sustainability perspective. The insights from viewpoint of engineering education practitioners are being highlighted.

3.2 Literature Review

The literature is analyzed from viewpoint of engineering education and sustainability:

Crofton (2000) aimed towards analyzing the path to enhance sustainable education among engineering students so that they can optimize the solutions for environment related problems. They suggested that engineering students and educators should have ability to adopt change and show interest towards sustainable education courses.

Abdul-Wahab et al. (2003) aimed towards highlighting the problems associated with environmental degradation and a need to provide education on sustainable concepts, tools and strategy among engineering students so as to create awareness among engineering students regarding sustainability issues. Their main aim was to educate students and make them to provide necessary solution for eliminating or minimizing environment-related problems.

Chisholm (2003) reviewed the critical factors having significant effect on sustainability in engineering education. They reviewed different curriculum models so as to enhance sustainability in engineering education. They did a comparison between traditional education models with off campus work based program. They found that work-based knowledge is more suitable for engineering students for their development and growth.

Alavi Moghaddam et al. (2007) presented the strategies of an Iran University adopted for sustainable development. They presented three strategies adopted by Iran University for successful deployment of sustainable goals in engineering education. They concluded with important drivers for achieving sustainable goals and they are student interest towards sustainable education, government encouragement, and faculty motivation and interest. They also presented the barriers and difficulties that the university faces towards achieving sustainability goals, they are: insufficient budget for deployment of strategies, resistance towards change among some people, and lack of experience towards sustainable development.

Apul and Philpott (2011) discussed sustainability based learning implemented in engineering education in University of Toledo and Fink's taxonomy of learning. They presented an example where engineering student of Toledo university worked on a project for supplying water to outdoor garden. They provided an engineering solution for the project. They also performed life cycle cost analysis and environmental impact assessment of the proposed solution. The authors suggested that with working on such project engineering student can develop their engineering and communication skills along with sustainability knowledge.

Arsat et al. (2011) presented three dimensions for characterizing sustainable courses in engineering education namely: model, orientations, and approaches. They reviewed 30 research papers and presented sustainable courses by characterizing them in three dimensions. They suggested that the mentioned three dimensions will help in designing sustainable courses for engineering students.

Watson et al. (2013) did a study on interest of engineering student towards sustainability. They performed a survey on 153 civil and environment engineering students of Georgia Tech University. They found that students were interested in gaining sustainability-related knowledge. They suggested that by enhancing the guidance level for implementation of sustainability in the design stage itself will lead to enhance engineering curriculum.

Hawkins et al. (2014) discussed the need for sustainability in engineering education. They also discussed the collaborative strategies of engineering institutes, companies and organizations for achieving sustainable goals. They have provided some suggestions to enhance engineering curriculum towards growth of engineering students for sustainable development.

Staniškis and Katiliūtė (2016) presented the result of QUESTE-SI (Quality system of European Scientific and Technical Education for Sustainable Industry) evaluation of selected program. They showed that a well-defined strategy is required for institutes to achieve sustainability environment in institutes. They have shown different ways by which an institute can become sustainable, they are linking problem to life cycle approach and considering social and economic aspect as well, providing sustainability related topics in related courses, enhancing partnership within members of institutes and among stakeholders.

Mulder (2017) discussed about strategic competencies for sustainable development of engineering education. They suggested that strategic competencies are very crucial for engineering student to provide incremental sustainable solution. They also specified that the biggest barrier towards sustainability is crucial curriculum of higher education which engage engineering student and restrict for other activities.

Esparragoza et al. (2018) presented an initiative taken by an academic institute to introduce sustainability in engineering curriculum to evaluate sustainability of product throughout its life span. The proposed approach will help in comparing sustainability performance of different alternatives to find the best alternative solution from sustainability viewpoint.

Thurer et al. (2018) presented a literature survey on integrating sustainability in engineering education. They considered 247 articles including 70 case studies. They discussed the results of integrating sustainability in engineering curriculum. They discussed the importance of stakeholders and strong political will towards sustainable development. They also presented the way sustainable education can be implemented in engineering education.

3.3 Methodology

The aim of this study is to systematically analyze the potential sustainability drivers for engineering education. A Multi Criteria Decision Making (MCDM) approach is implemented for prioritizing the drivers. Analytical Hierarchy Process (AHP) is a well-known and mostly used MCDM approach for prioritizing multiple criteria (Vinodh et al., 2011). AHP enables pairwise comparison of criteria and enablers by several experts to draw conclusion for prioritization of criteria and alternatives. The comparison can be made by giving absolute rating based on how much a factor is more important than other factor a particular attribute. The steps involved in AHP are described below:

Step 1: Collect and identify the list of drivers affecting sustainability in engineering education

Step 2: Collect the data from educational institute experts for pair wise comparison of drivers. The data can be collected in the form of rating that varies between 1 to 9 based on the relative importance.

For n driver a pairwise comparison matrix of n*n has shown below:

$$A = \begin{bmatrix} a_{11} & a_{12} & \cdots & a_{1n} \\ a_{21} & a_{22} & \cdots & a_{2n} \\ \vdots & \vdots & \ddots & \vdots \\ a_{n1} & a_{n2} & \cdots & a_{nn} \end{bmatrix}$$

Step 3: Normalize the pairwise comparison matrix so that all values in the matrix will lie in between 0 to 1. The normalized matrix is shown below:

$$A' = \begin{bmatrix} a'_{11} & a'_{12} & \cdots & a'_{1n} \\ a'_{21} & a'_{22} & \cdots & a'_{2n} \\ \vdots & \vdots & \ddots & \vdots \\ a'_{n1} & a'_{n2} & \cdots & a'_{nn} \end{bmatrix}$$

where

$$a'_{ij} = \frac{a_{ij}}{\sum_{i=1}^{n} a_{ij}} \quad \text{(Vinodh et al., 2011)} \tag{3.1}$$

i represents row and j represents column
n indicate number of drivers

Step 4: Calculate the eigen values and eigen vector from the normalized table

Formula for calculating eigen value is mentioned below:

$$W = \begin{bmatrix} W_1 \\ W_2 \\ \vdots \\ W_n \end{bmatrix}$$

where

$$W_i = \frac{\sum_{j=1}^{n} a'_{ij}}{n} \quad \text{(Vinodh et al., 2011)} \qquad (3.2)$$

W is the eigen vector and W_i is the eigen value for each drivers.

Step 5: Finally for prioritization of drivers, ranking need to be given to each driver based on their eigen value (Shankar et al., 2016).

Case Study

The aim of this study is to prioritize the drivers for sustainability in engineering education. From literature survey, 18 drivers have been identified influencing sustainability in engineering education. The identified drivers along with their description is mentioned in Table 3.1.

Step 2. Data is collected from educational experts for pairwise comparison between drivers. The data has been collected in the form of ratings which varies between 1 to 9.

Saaty scale has been used for pair wise comparison matrix. According to it, if two drivers A and B are in pair wise comparison then, if A is 4 times more important than B, it means for A to B rating 4 will be assigned. Similarly, if B is 4 times more important than A, it means for A to B rating 1/4 will be assigned. In a similar way, pair wise comparison matrix has been made. The collected data is shown in Table 3.2.

Step 3. The normalization matrix has been calculated in order to make each value lie in between 0–1. Equation (1) is used for normalization of matrix. The normalized matrix is shown in Table 3.3.

Step 4. The eigen value for each driver has been calculated using Equation 2.

The calculated eigen value of all 18 drivers are shown in Table 3.4.

Table 3.1 Identified drivers and their description

S. no.	Drivers	Description	Reference
1.	Declarative knowledge (D1)	Knowing the factual knowledge of the relevant discipline. i.e., real knowledge of required field.	Carew and Mitchell (2002)
2.	Theoretical knowledge (D2)	Knowing the principles and theories behind technology	Carew and Mitchell (2002)
3.	Critical thinking (D3)	Ability of an individual to take decision or to judge a value.	Carew and Mitchell (2002)
4.	Use of appropriate technology (D4)	Knowing about technology; appropriate technology for a particular process concerning sustainability	Crofton (2000)
5.	Inventive problem solving (D5)	Analyzing and solving problem by referring patterns of patents and inventions.	Byrne et al. (2013)
6.	Understanding of complexity and complex systems (D6)	Ability to analyze complex problem and provide necessary solution	Byrne et al. (2013)
7.	Deep understanding of sustainability concepts and tools (D7)	Should know concept of sustainability and have knowledge on sustainability assessment tools, Life Cycle Assessment (LCA) tools	Byrne et al. (2013)
8.	Quality of living (D8)	To provide proper education and practical knowledge to all so that they can gain experience and enhance their professional development and quality of living as well.	Valdes-Vasquez and Klotz (2010)
9.	Principles of engineering for sustainable development (D9)	To provide deep understanding of principles of engineering which can be used in making strategic plan to enhance sustainable performance.	García-Serna et al. (2007)

(Continued)

Table 3.1   (Continued)

S. no.	Drivers	Description	Reference
10.	DfE key strategies (D10)	It include environmental attributes into product in the design stage of product itself.	García-Serna et al. (2007)
11.	Sustainable design tools (D11)	To provide skills for sustainability concept and sustainable design tools for product development	El-Zein et al. (2008)
12.	Impact identification and developing alternative solutions to problems (D12)	To enhance the ability of student to identify the impact of project on environment and to provide necessary solutions.	El-Zein et al. (2008)
13.	Communication skills of students (D13)	To improve communication skill and personality development of student for professional growth	El-Zein et al. (2008)
14.	Improvement of student's skills in research (D14)	Enhancing skill in research can motivate students to utilize available resources and come up with good research.	El-Zein et al. (2008)
15.	Problem-based learning (D15)	In this learning, a student has been assigned some problems for analysis and solution derivation. From this kind of learning, a student can enhance problem solving capability.	Segalas et al. (2010)
16.	Self-learning (D16)	Self-learning is a desirable quality for student. Student should have self-learning ability so as to learn the engineering concept on own and try to implement such concepts in practical life.	Segalas et al. (2009)
17.	Cooperation and transdisciplinarity (D17)	Ability to communicate and cooperate with other departments, technical or non-technical departments for managing system	Segalas et al. (2009)
18.	Systemic thinking (D18)	Ability to analyse the problem by considering all facts and systematic thinking in all direction	Segalas et al. (2009)

Table 3.2 Pairwise comparison matrix

	Declarative knowledge	Theoretical knowledge	Critical thinking	Use of appropriate technology	Inventive problem solving	Understanding of complexity and complex systems	Deep understanding of sustainability concepts and tools	Quality of living	Principles of engineering for sustainable development	DfE key strategies	Sustainable design tools	Impact identification and developing alternative solutions to problems	Communication skills of students	Improvement of students skills in research	Problem-based learning	Self-learning	Cooperation and transdisciplinarity	Systemic thinking
Declarative knowledge	1	5	1/5	1/3	1/5	3	1/5	3	1/5	1/3	1	3	5	3	1/3	1/3	3	1/3
Theoretical knowledge	1/5	1	1/3	1/5	1/5	1/3	1/3	3	1/3	1/5	1/3	1/3	1	1/5	3	1/3	1/5	1/3
Critical thinking	5	3	1	3	1/3	3	1/3	3	1/5	1/5	1/3	1/3	3	1/3	3	3	3	1
Use of appropriate technology	3	5	1/3	1	1/3	1	1/3	3	1/5	1/5	1/3	1	3	1/3	1/3	1/3	1	1/3
Inventive problem solving	5	5	3	3	1	3	1	3	1/3	1/3	1/3	1/3	1	1/3	1	1	1	1
Understanding of complexity and complex systems	1/3	3	1/3	1	1/3	1	1/5	1/3	1/3	1/3	1/3	1/3	1	1	1	1/3	1/3	1
Deep understanding of sustainability concepts and tools	5	3	3	3	1	5	1	3	1/3	1	1	3	3	1/3	3	3	1	1
Quality of living	1/3	1/3	1/3	1/3	1/3	3	1/3	1	1/3	1/3	1/3	1/3	3	1/3	1/3	1/3	1	1/3
Principles of engineering for sustainable development	5	3	3	5	3	3	3	3	1	1	1	1	3	1/3	3	1/3	3	1/3

(Continued)

Table 3.2 (Continued)

	DfE key strategies	Sustainable design tools	Impact identification and developing alternative solutions to problems	Communication skills of students	Improvement of students skills in research	Problem-based learning	Self-learning	Cooperation and transdisciplinarity	Systemic thinking
Systemic thinking	1/3	3	1	1/3	1/3	1	3	1	1
Cooperation and transdisciplinarity	3	3	3	3	1	1	3	1	1
Self-learning	3	3	3	3	1/3	3	1	1/3	3
Problem-based learning	1	3	1	3	1/3	1	1	1/3	1
Improvement of students skills in research	1/3	1/3	1	3	1	1	1	1/3	1/3
Communication skills of students	3	3	5	1	3	3	3	1	3
Impact identification and developing alternative solutions to problems	3	3	1	5	1/5	1	1	1/3	1
Sustainable design tools	1	1	1/3	3	1/3	3	1/3	1/3	1/3
DfE key strategies	1	1	1/3	3	1/3	3	1	1/3	3
Principles of engineering for sustainable development	1	1	1	3	1/3	3	1/3	3	1/3
Quality of living	3	3	3	3	1/3	3	3	3	1
Deep understanding of sustainability concepts and tools	1	1	1/3	3	1/3	3	1/3	1/3	1
Understanding of complexity and complex systems	3	3	3	1	1	1	1	3	3
Inventive problem solving	3	3	1	3	1	3	1	1	1
Use of appropriate technology	5	3	3	1/3	3	3	3	1	3
Critical thinking	5	3	3	1/3	3	3	3	1	1
Theoretical knowledge	5	3	3	1	5	3	3	5	3
Declarative knowledge	3	1	1/3	1/5	1/3	3	3	1/3	3

Table 3.3 Normalization matrix

	D1	D2	D3	D4	D5	D6	D7	D8	D9	D10	D11	D12	D13	D14	D15	D16	D17	D18
D1	0.026	0.084	0.007	0.008	0.009	0.069	0.013	0.067	0.012	0.023	0.083	0.146	0.104	0.253	0.016	0.013	0.092	0.021
D2	0.005	0.017	0.011	0.005	0.009	0.008	0.022	0.067	0.020	0.014	0.028	0.016	0.021	0.017	0.016	0.013	0.006	0.021
D3	0.128	0.051	0.033	0.073	0.015	0.069	0.022	0.067	0.020	0.014	0.028	0.016	0.063	0.028	0.141	0.013	0.092	0.063
D4	0.077	0.084	0.011	0.024	0.015	0.023	0.022	0.067	0.012	0.014	0.028	0.016	0.063	0.028	0.016	0.013	0.031	0.021
D5	0.128	0.084	0.098	0.073	0.046	0.069	0.066	0.067	0.020	0.023	0.028	0.049	0.021	0.028	0.047	0.038	0.031	0.063
D6	0.009	0.051	0.011	0.024	0.015	0.023	0.013	0.007	0.020	0.023	0.028	0.016	0.021	0.084	0.047	0.013	0.010	0.021
D7	0.128	0.051	0.098	0.073	0.046	0.115	0.066	0.067	0.020	0.070	0.083	0.146	0.063	0.028	0.141	0.115	0.031	0.063
D8	0.009	0.006	0.011	0.008	0.015	0.069	0.022	0.022	0.020	0.023	0.028	0.016	0.063	0.028	0.016	0.013	0.031	0.021
D9	0.128	0.051	0.098	0.121	0.138	0.069	0.199	0.067	0.061	0.070	0.083	0.049	0.063	0.028	0.141	0.013	0.092	0.021
D10	0.077	0.084	0.164	0.121	0.138	0.069	0.066	0.067	0.061	0.070	0.083	0.146	0.063	0.028	0.047	0.115	0.092	0.021
D11	0.026	0.051	0.098	0.073	0.138	0.069	0.066	0.067	0.061	0.070	0.083	0.146	0.063	0.028	0.141	0.115	0.092	0.188
D12	0.009	0.051	0.098	0.073	0.046	0.069	0.022	0.067	0.061	0.023	0.028	0.049	0.104	0.084	0.047	0.115	0.092	0.063
D13	0.005	0.017	0.011	0.008	0.046	0.023	0.022	0.007	0.020	0.023	0.028	0.010	0.021	0.028	0.016	0.013	0.031	0.021
D14	0.009	0.084	0.098	0.073	0.138	0.023	0.199	0.067	0.183	0.210	0.250	0.049	0.063	0.084	0.047	0.115	0.031	0.188
D15	0.077	0.051	0.011	0.073	0.046	0.023	0.022	0.067	0.020	0.070	0.028	0.049	0.063	0.084	0.047	0.115	0.031	0.063
D16	0.077	0.051	0.098	0.073	0.046	0.069	0.022	0.067	0.183	0.023	0.028	0.016	0.063	0.028	0.016	0.038	0.092	0.021
D17	0.009	0.084	0.011	0.024	0.046	0.069	0.066	0.022	0.020	0.023	0.028	0.016	0.021	0.084	0.016	0.013	0.031	0.063
D18	0.077	0.051	0.033	0.073	0.046	0.069	0.066	0.067	0.183	0.210	0.028	0.049	0.063	0.028	0.047	0.115	0.031	0.063

Table 3.4   Eigen value of drivers

Drivers	Eigen Value (W)
Declarative knowledge (D1)	0.05816
Theoretical knowledge (D2)	0.017522
Critical thinking (D3)	0.051952
Use of appropriate technology (D4)	0.031378
Inventive problem solving (D5)	0.054434
Understanding of complexity and complex systems (D6)	0.02428
Deep understanding of sustainability concepts and tools (D7)	0.078017
Quality of living (D8)	0.023364
Principles of engineering for sustainable development (D9)	0.082887
DfE key strategies (D10)	0.08408
Sustainable design tools (D11)	0.087495
Impact identification and developing alternative solutions to problems (D12)	0.061164
Communication skills of students (D13)	0.019434
Improvement of student's skills in research (D14)	0.106138
Problem-based learning (D15)	0.055564
Self-learning (D16)	0.056167
Cooperation and transdisciplinarity (D17)	0.035917
Systemic thinking (D18)	0.072084

3.4 Results

The main aim of this study is to prioritize the drivers of sustainability in engineering education. 18 drivers have been collected from literature survey and has been analyzed using AHP approach. The prioritization ranking is shown in Table 3.5.

From Table 3.5, it can be seen that 'Improvement of student's skills in research (D14)' is the top most priority driver with eigen value of 0.106 for sustainability in engineering education. The second top most priority driver is 'Sustainable design tools (D11)' with eigen value of 0.0875. The driver which is having least importance on sustainability in engineering education is 'Theoretical knowledge (D2)' with eigen value of 0.0175. The ranking order is D14> D11> D10> D9> D7> D18> D12> D1> D16> D15> D5> D3> D17> D4> D6> D8> D13> D2. The priority order of drivers is discussed further with experts of engineering education and their views are in agreement with the generated order of drivers.

Table 3.5 Rank priority of drivers

Drivers	Rank
Declarative knowledge (D1)	8
Theoretical knowledge (D2)	18
Critical thinking (D3)	12
Use of appropriate technology (D4)	14
Inventive problem solving (D5)	11
Understanding of complexity and complex systems (D6)	15
Deep understanding of sustainability concepts and tools (D7)	5
Quality of living (D8)	16
Principles of engineering for sustainable development (D9)	4
DfE key strategies (D10)	3
Sustainable design tools (D11)	2
Impact identification and developing alternative solutions to problems (D12)	7
Communication skills of students (D13)	17
Improvement of student's skills in research (D14)	1
Problem-based learning (D15)	10
Self-learning (D16)	9
Cooperation and transdisciplinarity (D17)	13
Systemic thinking (D18)	6

3.5 Conclusions

During the recent time, sustainable concepts are being discussed in the context of engineering education. The key reasons include: industry demands for sustainable product development, to fulfill government norms, to increase awareness among engineering student, to understand sustainable education concepts and application, and Environmental degradation (Glavic, 2006; Hawkins et al., 2014; Abdul-Wahab et al., 2003). In order to enable sustainability in engineering education perspective, drivers need to be identified and prioritized. This chapter presents the analysis of 18 drivers from literature. The priority order of driver is obtained and top key drivers include 'Improvement of student's skills in research (D14)', 'Sustainable design tools (D11)', and 'DfE key strategies (D10)'. The opinion of experts in engineering education also confirmed the priority order. In future, additional drivers from viewpoint of advances in engineering education could be included.

References

Abdul-Wahab, S. A., Abdulraheem, M. Y., and Hutchinson, M. (2003). The need for inclusion of environmental education in undergraduate engineering curricula. *International Journal of Sustainability in Higher Education, 4(2),* 126–137.

Alavi Moghaddam, M. R., Taher-shamsi, A., and Maknoun, R. (2007). The role of environmental engineering education in sustainable development in Iran: AUT experience. *International Journal of Sustainability in Higher Education, 8(2),* 123–130.

Apul, D. S., and Philpott, S. M. (2011). Use of outdoor living spaces and Fink's taxonomy of significant learning in sustainability engineering education. *Journal of professional issues in engineering education and practice, 137*(2), 69–77.

Arsat, M., Holgaard, J. E., and de Graaff, E. (2011). Three dimensions of characterizing courses for sustainability in engineering education: Models, approaches and orientations. In *Engineering Education (ICEED), 2011 3rd International Congress on* (pp. 37–42). IEEE.

Byrne, E. P., Desha, C. J., Fitzpatrick, J. J., and "Charlie" Hargroves, K. (2013). Exploring sustainability themes in engineering accreditation and curricula. *International Journal of Sustainability in Higher Education, 14*(4), 384–403.

Carew, A. L., and Mitchell, C. A. (2002). Characterizing undergraduate engineering students' understanding of sustainability. *European journal of engineering education, 27*(4), 349–361.

Chisholm, C. U. (2003). Critical factors relating to the future sustainability of engineering education. *Global J. of Engng. Educ, 7*(1), 29–38.

Crofton, F. S. (2000). Educating for sustainability: opportunities in undergraduate engineering. *Journal of Cleaner Production, 8*(5), 397–405.

El-Zein, A., Airey, D., Bowden, P., and Clarkeburn, H. (2008). Sustainability and ethics as decision-making paradigms in engineering curricula. *International Journal of Sustainability in Higher Education, 9*(2), 170–182.

Esparragoza, I. E., Mesa, J. A., and Maury, H. E. (2018). Introducing sustainability in engineering design education: a case study using analysis of impacts during the design for sustainability (AID-DS). In *DS92: Proceedings of the DESIGN 2018 15th International Design Conference* (pp. 2429–2440).

García-Serna, J., Pérez-Barrigón, L., and Cocero, M. J. (2007). New trends for design towards sustainability in chemical engineering: Green engineering. *Chemical Engineering Journal, 133*(1–3), 7–30.

Glavic, P. (2006). Sustainability engineering education. *Clean Technologies and Environmental Policy, 8*(1), 24–30.

Hallstedt, S. I. (2017). Sustainability criteria and sustainability compliance index for decision support in product development. *Journal of Cleaner production, 140*, 251–266.

Hawkins, N. C., Patterson, R. W., Mogge, J., and Yosie, T. F. (2014). Building a sustainability road map for engineering education. *ACS sustainable chemistry,* 340–343.

Mulder, K. F. (2017). Strategic competences for concrete action towards sustainability: An oxymoron? Engineering education for a sustainable future. *Renewable and Sustainable Energy Reviews, 68*, 1106–1111.

Segalàs, J., Ferrer-Balas, D., and Mulder, K. F. (2010). What do engineering students learn in sustainability courses? The effect of the pedagogical approach. *Journal of Cleaner Production, 18*(3), 275–284.

Segalàs, J., Ferrer-Balas, D., Svanström, M., Lundqvist, U., and Mulder, K. F. (2009). What has to be learnt for sustainability? A comparison of bachelor engineering education competences at three European universities. *Sustainability Science, 4*(1), 17.

Shankar, K. M., Kumar, P. U., and Kannan, D. (2016). Analyzing the drivers of advanced sustainable manufacturing system using AHP approach. *Sustainability, 8*(8), 824.

Staniškis, J. K., and Katiliūtė, E. (2016). Complex evaluation of sustainability in engineering education: case and analysis. *Journal of Cleaner Production, 120*, 13–20.

Thürer, M., Tomašević, I., Stevenson, M., Qu, T., and Huisingh, D. (2018). A systematic review of the literature on integrating sustainability into engineering curricula. *Journal of Cleaner Production, 181*, 608–617.

Valdes-Vasquez, R., and Klotz, L. (2010). Incorporating the social dimension of sustainability into civil engineering education. *Journal of Professional Issues in Engineering Education & Practice, 137*(4), 189–197.

Vinodh, S., Shivraman, K. R., and Viswesh, S. (2011). AHP-based lean concept selection in a manufacturing organization. *Journal of Manufacturing Technology Management, 23*(1), 124–136.

Watson, M. K., Noyes, C., and Rodgers, M. O. (2013). Student perceptions of sustainability education in civil and environmental engineering at the Georgia Institute of Technology. *Journal of Professional Issues in Engineering Education and Practice, 139*(3), 235–243.

4

Visualization Technologies in Construction Education: A Comprehensive Review of Recent Advances

Akeem Pedro[1], Rahat Hussain[1], Anh-Tuan Pham-Hang[2] and Hai Chien Pham[3*]

[1]Department of Architectural Engineering, Chung-Ang University, Seoul 06974, South Korea
[2]School of Computer Science and Engineering, International University – Vietnam National University HCMC, Ho Chi Minh City 7000000, Vietnam
[3]Faculty of Civil Engineering, Ton Duc Thang University, Ho Chi Minh City 7000000, Vietnam
E-mail: lanrepedro3@gmail.com; rahat4hussain@gmail.com; anhtuanphamhang@gmail.com; phamhaichien@tdtu.edu.vn;
*Corresponding Author

Over the past few decades, visualization technologies such as Virtual Reality (VR) and Augmented Reality (AR) and Building Information Modeling (BIM) have advanced rapidly, presenting unprecedented opportunities and solutions for the construction industry. Considerable research efforts have focused on innovatively applying advanced visualization techniques to address various challenges in construction education. In this regard, this paper comprehensively synthesizes research efforts on visualization technologies in construction education with a systematic review of studies from 2005 to 2016. Through a systematic search strategy, 108 papers are found and considered relevant to the research based on defined selection criteria. In order to summarize past research works, identify issues, gaps, and areas of concentration a four layer architecture which classifies studies based on the concept and theory, development, evaluation and adoption layers is applied. This paper analyses past research efforts on the application of advanced visualization technologies in construction education, and then identifies trends, deficiencies and opportunities. The comprehensive review and analyses presented in this paper would be beneficial to researchers and practitioners, serving as a foundation to pave the way for future research on advanced visualization technologies in construction education.

4.1 Introduction

Over the past few decades, the construction industry has advanced tremendously, embracing cutting-edge technologies with great benefits through various construction stages. Given the complex and dynamic nature of construction, advanced visualization techniques such as BIM, Virtual Reality (VR) and Augmented Reality (AR) have been utilized to support visual communication, and make various construction processes more efficient, effective and safe. Numerous industry-oriented studies have adopted visualization technologies for enhancing safety and health management (Sacks and Pikas, 2013; Hussain et al., 2018); real-time jobsite monitoring (Teizer et al., 2010; Hussain et al., 2017); and for developing automated planning systems (Zhang et al., 2013). Likewise, the potential of these technologies has been considered for tertiary education in various Architecture Engineering and Construction (AEC) disciplines. Conventional teacher-centered didactic methods in the AEC domains have been noted to limit student engagement, motivation and understanding (Pedro et al., 2016; Pham et al., 2018; Park et al., 2016). Hence, in order to improve construction pedagogy, a plethora of studies have focused on visualization technologies to equip learners with more visual, interactive and captivating learning contents and activities.

Several articles have reviewed recent advances in virtual reality, augmented reality, simulation games, and BIM for the construction industry. Studies have conducted reviews focusing on summarizing and analyzing research on: visualization (Bouchlaghem et al., 2005); digital design (Zhou et al., 2012); virtual reality for safety (Bhoir and Esmaeili, 2015); simulation games (Deshpande and Huang, 2011) and augmented reality (Chi et al., 2013; Rankohi and Waugh, 2013; Wang et al., 2013) in the AEC domains. However, aside from a state of the art review of virtual reality in built environment education (Keenaghan and Horváth, 2014), research efforts thus far have not comprehensively reviewed or synthesized literature on visualization technologies in construction education. As visualization technologies consistently become more pervasive and commonplace in society, research which consolidates extant knowledge, achievements and trends on their application would be beneficial to pave the way for future research in the construction education. Hence, this study aggregates 108 research studies and conducts a systematic review from 2005 to 2016, with the goal of comprehensively synthesizing research in the field. The aims of the review are to: shed light on the nature and extent of visualization technology applications in construction education; identify the major areas of research concentration and deficiency; and take stock of research trends in literature over the years. To this end, a four layer architecture (comprising of the concept, implementation, evaluation and adoption layers) is applied in order to extract information from studies and enable the classification and analysis of literature. This article contributes to scientific research through a comprehensive review which provides a broad overview visualization technologies, and their level of maturity in construction education. The discussions presented show researchers and practitioners how visualization technologies have been applied to address various challenges in construction education, and suggest how they can be applied in future research.

The paper is structured as follows. After the introduction, the current state of visualization technologies in construction education is discussed in detail. Prominent techniques such as Virtual Reality, Augmented Reality and BIM are introduced and their significance for construction education is discussed. Next, the research methodology expounds on the scope of the review and the selection criteria for articles in Section 3. Selected papers are classified and analyzed based on a 4 layer architecture in terms of: (1) concept and theory, (2) Implementation, (3) Evaluation, and (4) Adoption in Section 4. Section 5 presents the discussion and summary section, which highlights main insights, emerging research issues, gaps and potential future research directions. Lastly, the conclusion section summarizes and concludes the paper.

4.2 Construction Education and the Need for Enhanced Visualization

This section discusses recent research works relating to the state of construction education and the need for enhanced visualization in pedagogy. It defines the prominent visualization technologies which form the basis for the review presented in this paper.

4.2.1 Need for Visualization Support in Construction Education

Construction is described as a distinctly unique and highly fragmented industry, with work environments where various factors are involved and countless activities occur simultaneously (Le et al., 2015). Tertiary construction education plays a crucial role in equipping students with the knowledge, skills and competencies necessary for effectively working in the construction industry. Construction graduates are required to have thorough knowledge and understanding of various subjects including construction design, management, materials, equipment, methods, scheduling, structures, sustainability and so on (Pedro et al., 2018). Traditionally, learners receive instruction through teacher-centered lectures in classroom environments (Le et al., 2015). However, traditional instructional models can be problematic (Shirazi and Behzadan, 2015), as lecture teaching styles sometimes falls short in serving as an effective communication tool for transferring knowledge to learners (Irizarry et al., 2012). Occasionally, learners also have opportunities to go on field trips, intended to provide a more realistic understanding of construction (Park et al., 2016; Shirazi and Behzadan, 2015). However, numerous limitations prevent learners from reaching their learning goals with both conventional lectures and field trips. As mentioned by Forsythe (Forsythe, 2009), a conundrum exists between construction principles taught in the classroom and a practical understanding of what actually happens on the jobsite.

Construction engineering is recognized as one of the disciplines in which learners have the most difficulty in visualizing and understanding complex and abstract information (Nikolic et al., 2011). A disconnection from the realistic context is one of the main challenges hindering the effectiveness of traditional instructional models in construction education (Becerik-Gerber et al., 2012), as paper based approaches fail to provide a realistic sense of the challenges encountered on site (Liaw et al., 2012), and students cannot gain exposure to the complex issues related to construction processes (Shanbari et al., 2016). Furthermore, numerous subjects in construction require learners to develop 3D models mentally by visualizing the different components of building structures. Learners without previous practical experience may lack the ability to fully visualize details of structures (Vimonsatit and Htut, 2016), and may face challenges and require more time in developing visual 3D models (Irizarry et al., 2012).

In addition, as noted by Perdomo et al. (2005) and Park et al. (2016) difficulties are encountered in portraying complex details with conventional teaching tools in construction education. Difficulties in perceiving contextual details may restrict learner's abilities to understand spatial constraints with building elements (Shanbari et al., 2016). Learners may also struggle to comprehend complex designs with conventional educational approaches

(Wong et al., 2011). In the case of construction scheduling, conventional teaching methods use bar charts and network diagrams which may portray an unclear picture, leading to a flawed understanding of the construction processes (Haque et al., 2007).

Construction courses typically use images and videos to assist learners in visualizing various construction elements, tasks, processes and environments. However, learners cannot interact with these contents; hence they often play a passive role in the classroom. The use of advanced visualization technologies in conjunction with interactive features, simulation and game-play dynamics has been shown to afford captivating and engaging pedagogic contents (Pedro et al., 2016; Haque et al., 2007).

Even though field trips present an opportunity to address the aforementioned issues, difficulties with scheduling site visits, ensuring students' safety and acquiring site access are often encountered. Moreover, it is difficult to organize many field trips related to specific on-going course topics (Liaw et al., 2012; Goedert et al., 2011). As a result, graduates may enter the construction industry without sufficient experience and knowledge (Pham et al., 2017, 2018). There is a need for visualization to addresses the educational challenges students face in construction (Irizarry et al., 2012).

4.2.2 Prominent Visualization Techniques in Construction Education

In order to address the aforementioned issues and enhance learning processes, scholarly efforts have considered the use of visualization techniques and approaches such as Virtual Reality and environments, Augmented Reality and BIM. Studies have demonstrated that visualization techniques can support experiential learning (Haque et al., 2007), improve understanding of spatial configurations (Sampaio et al., 2009), provide a sense of scales and dimensions of structures (Vimonsatit and Htut, 2016) and also enhance students' abilities in conceptualizing and comprehending construction concepts (Clevenger et al., 2012).

VR refers to computer generated artificial environments which are experienced through sensory stimuli and can be interacted with in a seemingly realistic way. A few main VR display technologies have emerged over the years. On computers, VR is primarily experienced through two of the five senses, namely sight and sound. The use of large projection screens (power-walls), CAVE Automatic Virtual Environments (CAVEs) and Head Mounted Displays (HMDs) is also possible. With the consistent advances in mobile computing and visualization technologies, VR-glasses which take advantage of the processing power, high definition display, location-tracking and built in functions of smart devices have also emerged.

The medical, engineering, and manufacturing disciplines have successfully used VR to create innovative learning environments that simulate realistic physical spaces (Aurich et al., 2009). Similarly, construction education has had its share of VR systems and environments with applications in building construction (Sampaio and Martins, 2014), construction safety (Lin et al., 2011), equipment training (Guo et al., 2012) and so on. Virtual environments are defined as the visual stimuli, and optionally other senses, presented by virtual and augmented reality systems (Rebenitsch and Owen, 2016). In tertiary institutions, studies have advocated VR and web based virtual environments such as second life to support learner collaboration (Haque et al., 2007; Le et al., 2015), support detailed visualization of complex elements (Park et al., 2016), improve captivation and engagement (Pedro et al., 2016), and improve learning outcomes in various aspects of construction education.

Augmented Reality, which creates environments where digital information is overlaid onto a predominantly real world view has been used to address visualization difficulties in various areas and domains such as engineering, military and automotive industries. AR gives users a view of the real world where elements are superimposed by computer generated contents such as graphics, sounds, videos, or digital information (Rankohi and Waugh, 2013). AR has also garnered attention in construction over recent years, with majority of applications focusing on addressing issues in actual industry practice. For example, AR based systems have been developed for onsite safety management (Park and Kim, 2013), facility management (Gheisari et al., 2013), and defect management (Kwon et al., 2014). Similarly, AR has been used to present information and enhance learning in construction pedagogy, with applications for construction equipment (Behzadan and Kamat, 2013) and safety courses (Le et al., 2015). AR can motivate students to learn, and also help students to visualize virtual information in the context of a specific physical space (Ayer et al., 2016). In addition to these benefits, adopting advanced visualization technologies with e-learning approaches can facilitate self-paced learning and distance learning, allowing large audiences to access virtual learning contents via the web (Wong et al., 2011; Ayer et al., 2016).

Aside from VR and AR, BIM has also played a revolutionary role in providing visualization support in the construction industry. BIM is formally defined as the digital representation of the physical and functional characteristics of a facility. Rather than a mere technique, BIM is considered both a technology and a process. The technology aspect of BIM allows project stakeholders visualize construction works within simulated environments to identify potential risks or issues. In terms of visualization, BIM can portray complex details of building components, provide realistic perceptions of buildings, support the consideration of various alternatives and aid in clash detection. Rather than static visualizations, BIM can be used to illustrate construction work sequencing, and safe work strategies (Clevenger et al., 2012). In reinforced concrete design, 3D visualization of concrete members can advance students' understanding of reinforcement details and rebar-placement (Irizarry et al., 2012).

The process aspect refers to BIM as an organizational solution which affords close communication, collaboration and the integration of diverse stakeholder roles on a project (Sattineni et al., 2012). After reshaping the Architecture Engineering and Construction (AEC) industries, BIM applications have also extended into tertiary construction education. BIM has been introduced not only as a tool for improving visualization in existing construction courses, but also as a freestanding course with its own requirements in tertiary institutions (Sacks and Pikas, 2013).

Over the last decade, there has been consistent research on the application of visualization technologies in various facets of construction education. In order to clearly establish the current state of visualization technologies in construction education, it is now necessary to conduct a comprehensive review, to synthesize research efforts thus far. This would be beneficial in identifying research achievements, gaps and guiding future research towards improvements and greater levels of achievement in construction pedagogy.

4.3 Review Methodology

In order to comprehensively synthesize research on the application of visualization technologies in construction education, a systematic review process was followed. A systematic review seeks to systematically search for, appraise, and exhaustively synthesize research evidence, to demonstrate what is known, what remains unknown, and what should be considered for future research (Grant and Booth, 2009). The main objectives of this review are to:

- Summarize existing evidence on the nature and extent of visualization technology applications

- Reveal areas of concentration and identify any gaps and deficiencies in the field
- Identify trends and suggest areas for further research

4.3.1 Search Strategy

This paper presents a review of research efforts concerning visualization technologies for tertiary construction education in the last decade, from 2005 to 2016. A thorough literature search strategy comprising of several distinct steps was defined in order to exhaustively identify eligible articles for this study. The first step involved searching multi-disciplinary digital databases and search engines applicable to fields such as engineering, education, information science, technology and so on. The main databases selected for this purpose were Scopus and Google Scholar. Scopus was chosen as a primary source of research evidence for two reasons: firstly, because it is recognized as one of the most extensive databases in the scientific community; and secondly because it covers a wider range of journals and more recent publications than other online bibliometric sources (Chadegani et al., 2013). Secondly, the Google Scholar was selected as the other main source of articles because of its broad content coverage, and indexing of thousands of academic journals (Jacso, 2008). Aside from electronic databases, the internet can also be considered as an addi tional source of evidence in a review (Kitchenham, 2004). This point further motivated the selection of Google Scholar to improve the comprehensiveness of the review, since it is essentially an internet search engine. The second step of the search strategy involved searching the following databases: Web of Science, Jstor, ACM Digital Library, IEEE Explore and Ingenta Journals. Next, online databases of construction journals such as Automation in Construction, Journal of Information Technology in Construction, Computing in Civil Engineering, International Journal of Construction Education and Research were searched. The aforementioned are considered reputable and leading journals in the AEC domains (Wang et al., 2013; Kitchenham, 2004). Lastly, the snowball method was used to search for more studies from the references of all eligible papers.

4.3.2 Selection of Articles for Inclusion

After a list of potentially relevant studies was compiled, actual relevance was assessed based on the following four eligibility criteria:

1. Only publication dates ranging from 2005 to 2016, inclusive were considered in order to limit the scope and offer a recent perspective in the review.
2. Application of visualization technologies: Selected papers were required to pertain to visualization technologies such as virtual reality and environments, augmented reality, simulation games and BIM. Papers which did not apply any visualization technologies were excluded.
3. In construction education: All papers were required to pertain to construction education. Hence, papers that focused on the application of visualization technologies in construction jobsites and industry practice were excluded.
4. Specifically for visualization purposes: Selected studies were required to apply visualization technologies specifically for visualization purposes. It was essential to ascertain this for technologies such as BIM, which can be regarded in diverse ways, as a course, process and a technology. Hence, studies which did not consider BIM models or environments for visualization purposes were excluded.

Initially, relevant journal papers and conference proceedings were retrieved through a keyword search with the aforementioned online databases and search engines. The main keywords used were "virtual", "virtual reality", "visualization", "BIM", "3D", "4D", "CAD (Computer Aided Design)", "augmented reality", "simulation", and "game", "web 3D" in "construction education" and "built environment education". As depicted in Figure 4.1, the

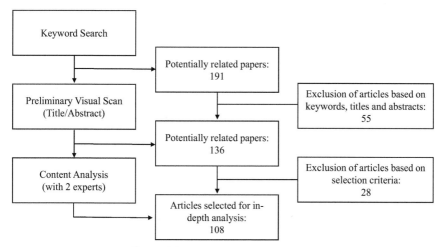

Figure 4.1   Paper selection procedures.

initial keyword searches resulted in a total of 191 potentially related research articles. After further review of paper titles, abstracts and keywords, the list of papers was narrowed down to 136 papers. Finally, after full texts had been retrieved and their contents reviewed, 108 papers were eligible for the study based on the aforementioned selection criteria. Furthermore, in order to avoid bias, final inclusion decisions were made in consultation with two experts, and consensus was required on all decisions.

4.4 Review of Visualization Technologies in Construction Education

This section presents the review of 108 research papers pertaining to visualization technologies in construction education. Selected papers are classified according to their, respective journals or conferences, years of publication technological focus, and area of application; and then analyzed in further detail according to concept and theory, implementation, evaluation and adoption based on the four layer classification architecture.

4.4.1 Classification of Articles based on Journals and Conferences

The classification of articles according to journals and conferences is presented in Table 4.1. As illustrated in the table, most research works on visualization technologies in construction education were from proceedings of the American Society for Engineering Education conferences (18 articles, 16.67%). ASEE is a non-profit professional organization recognized as a leader in collecting facts on engineering education, and promoting the role of engineering and engineering technology education (ASEE, 2017). In terms of journals, the Journal of Information Technology in Construction published the most articles on visualization technologies in construction education (12 articles, 11.11%), followed by the Journal of Professional Issues in Engineering Education with (8 papers, 7.4%), and the International Journal of Construction Education and Research (5 papers, 4.63%). Among the remaining articles, some were published in technology-oriented AEC journals (such as Automation in construction, Computing in Civil Engineering), while others were spread across built environment, education and visualization related journals and conferences.

Over the past decade research on visualization technologies in construction education has extended into a variety of disciplines, with significant works in IT-oriented, education-oriented and management

Table 4.1   Classification of articles based on conferences and journals in order of papers

Journal and Conference List	Number of Papers
American Society for Engineering Education Conference Proceedings	18
Journal of Information Technology in Construction	12
Journal of Professional Issues in Engineering Education and Practice	8
IEEE Conference Proceedings	7
Construction Research Congress Proceedings	6
International Journal of Construction Education and Research	5
Architectural Engineering Institute Conference Proceedings	3
Automation in Construction	3
Computing in Civil and Building Engineering	3
Computing in Civil Engineering	3
Journal of Construction Engineering and Management	3
Winter Simulation Conference	3
Advances in Engineering Education	2
International Conference Computer Graphics, Visualization, Computer Vision and Image Processing	2
International Journal of Engineering Education	2
Journal of Computing in Civil Engineering	2
Others	26
Total	**108**

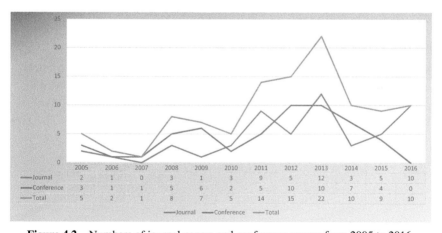

Figure 4.2   Numbers of journal papers and conference papers from 2005 to 2016.

oriented journals and conferences. As illustrated in Figure 4.2, majority of research on the field has been published in conference papers and proceedings over the years, however from 2014, publications have emerged predominantly from journals.

4.4.2 Classification of Articles based on Publication Year and Technology

Figure 4.3 illustrates the number of articles concerning applications of visualization technologies in construction education from 2005 to 2016 according to publication year, and technology type. Visualization approaches are broken down into four main types, namely BIM, AR, VR, nD (which include 3D and 4D visualization technologies), and others. Majority of the studies included in the review were published after 2010 (85 articles, 78.7%). As illustrated in the figure, there was a spike in visualization technology related research from 2010 to 2013, with cumulative research efforts increasing rapidly. However, 2013 and 2014 saw a major decline in research numbers. From 2014 to 2016 research on visualization technologies appears steady, without any major peaks or drops. It also becomes apparent that BIM is the most prominent visualization technology, with a stable increase in research from 2010 to 2014. This is not surprising considering the fact that BIM is a visualization approach which originated specifically for information management and visual communication in the construction domain. On the other hand, the decline of research on the application of 3D and 4D visualization between 2009 and 2012 suggests that these technologies are no longer of interest in construction education research.

4.4.3 Articles by Area of Application

Articles were classified according to their application areas in construction education over the 11 years period analyzed. As illustrated in Figure 4.4, majority of the articles focused on construction education in general (25 articles, 23.12%), and construction management (23 articles, 21.3 %). These

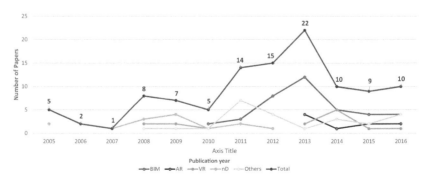

Figure 4.3   Classification of articles based on publication year and technology type.

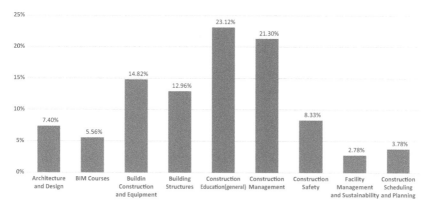

Figure 4.4 Classification of articles in each application area.

are followed by studies focused on building construction and equipment (16 articles, 14.82%), with papers adopting visualization techniques for education on construction earthwork (Duckworth et al., 2012), construction process simulation (Forsythe, 2009; Rebenitsch and Owen, 2016; Duckworth et al., 2012) and masonry (Shanbari et al., 2016). Next, 12.96% (14 articles) focus on visualization technique for building structures. Studies in this area focus on structural design, steel structures, steel connections and bridges. Recently, several studies have also explored the potentials of visualization technologies for construction safety education (9 articles, 8.33%). Considering the poor safety record of the construction industry, safety education presents an opportunity to enhance the safety knowledge and performance of future construction personnel. In the context of construction safety, one of the main benefits of virtual environments is that they allow users to perceive the consequences of their decisions and actions in a safe environment, without detrimental consequences in real life (Sacks and Pikas, 2013). Lastly, the areas of visualization technology application with the least research are architecture and design (8 articles, 7.40%), BIM courses (6 articles, 5.56%), construction scheduling and planning (4 articles, 3.78%) and facility management and sustainability (3 articles, 2.78%).

4.4.4 Classification and Review based on Four Layer Architecture

Based on an adaptation of the standard architecture layer for emerging information technology applications (Wang et al., 2013; Kitchenham, 2004; Nikolic et al., 2009), pertinent literature were classified in 4 layers, namely:

Table 4.2   Number and percentage of papers by classification area

Classification Criteria	Number of Articles	Global Percentage of Subject (%)
1. Concept and Theory	56	51,85
2. Implementation	51	47,22
3. Evaluation	53	49,07
4. Adoption	21	19,44
Total	**108**	**100**

concept and theory, implementation, evaluation and adoption. Many articles fit into multiple categories, hence they were included in several categories. As a result, the sum of articles in Table 4.2 is greater than the total (108 articles). In order to prevent bias and subjective classification, the categorization of papers was carried out with two expert researchers. Following in depth analysis, consensus was reached regarding the classification of each article.

Among the studies included in the review, majority pertained to the concept and theory area (56 articles), closely followed by 53 articles involving the assessment and evaluation of visualization technology based systems. Similarly, the implementation layer received significant attention with 51 papers. Lastly, the adoption layer received the least amount of attention, with only 21 articles.

4.4.4.1 Concept and theory layer

The concept and theory layer refers to studies which present novel ideas, theories, frameworks or concepts regarding how visualization technologies can be developed, applied, evaluated and adopted within the scope of construction education. These include conceptual frameworks, giving users an idea of the detailed underpinnings of a novel system; development and implementation frameworks, considering how to actualize visualization based education systems; and evaluation and adoption frameworks, which explore how to assess innovative systems and incorporate them into construction curricula. As shown in Table 4.2, majority of studies in the review (56 articles, 53.48%) were related to the concept and theory layer. Regarding the visualization techniques considered, most studies related to VR and Virtual World based systems (27 articles, 48.21%), followed by BIM based systems with 20 articles (35.71%). Only 3.57% of articles related to the application of augmented reality for construction education. This is not surprising considering the fact that the main benefit of AR is through its effectiveness in providing convenient access to information in the context where it is needed, e.g. on the construction site during construction work execution.

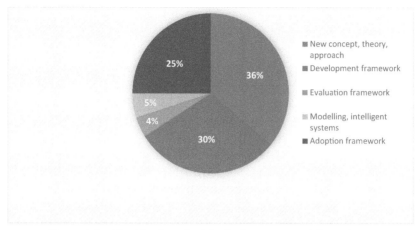

Figure 4.5 Classification of articles in the concept and theory layer.

Articles in this layer were categorized according five groups. As illustrated in Figure 4.5, these include: new concepts, theories and approaches, development frameworks, evaluation frameworks, modelling and intelligent systems, and adoption frameworks. Most articles in this layer fit into the new concept, theory or approach category. (20 articles, 36%). For example, tele present augmented reality for discovery-based construction education, and social virtual reality for construction safety (Le et al., 2015). Aside from applying visualization technologies in specific aspects of construction education, a few studies also developed entirely new innovative theories and approaches to construction education. Noteworthy approaches include: Building Interactive Modeling (BiM), which represented a vision of combining virtual worlds and BIM for virtual collaboration and learning (Ku and Mahabaleshwarkar, 2011); Virtual Interactive Construction Education (VICE), and Interactive Building anatomy Modeling (IBAM) (Park et al., 2016), which combined medical anatomy theory with the interactive features of VR to afford enhanced building construction education.

The new concept, theory and approach category was followed by development frameworks (17 articles, 30%) and adoption frameworks (14 articles, 25%). Modelling and intelligent systems constituted 5% (3 articles) respectively. These included (Obonyo, 2011) proposing an application that combined advanced visualization and interactive management through complex virtual devices and intelligent components; and (Goedert et al., 2011) with a construction education framework based on deductive synthesis and automated inference technologies. However, aside from these studies, research efforts have not considered the development of modelling and intelligent systems alongside visualization technologies in construction education. More attention paid to this area could make the proposition of technology-assisted systems more attractive. Lastly, evaluation frameworks received the least attention with 4% of this layer (2 articles).

4.4.4.2 Implementation layer
The implementation layer refers to studies which developed visualization based contents or systems utilizing VR, AR, BIM, 3D, 4D visualizations or simulation games. It includes hardware and software, content and interaction designs, as well as agent-based and knowledge-based visualization systems. As depicted in Table 4.2, almost half of the articles included in this review (51 articles, 47.22%) involved the development and implementation of visualization based systems. Most of the implemented systems applied VR as their main visualization approach (31 articles, 60.78%), which was much greater than applications of AR (7 articles, 13.72%), and BIM (5 articles, 9.8).

Visualization technologies deployed in this layer were categorized according to their levels of technology maturity. After reviewing all implementations, articles fit into the following categories: 1) proof of concepts; which include a preliminary example or scenario; 2) small scale prototypes, which involved a degree of system development and a few demonstrative prototype scenarios; and 3) full systems, which applied fully functional visualization systems, tools or applications. As illustrated in Figure 4.6, majority of visualization technology implementations in construction education reached the small scale prototype maturity level (52.95%), while 39.21% have created proof of concepts. However, the development of full systems has been rare (7.84%).

Figure 4.6   Maturity of visualization technology implementations in construction education.

25.49% of articles in this layer (13 articles) based their systems developments on pedagogic learning theories. Prominent ones included collaborative learning (6 articles), experiential learning (4 articles), game-based, interoperable and discovery-based learning. Most studies in this layer dealt with content design (49 articles, 96.07%), while 2 articles (3.92%) deal with intelligent, agent-related visualization systems. Most articles proposed single user visualization supported systems (40 articles, 78.43%), while 11 articles (21.57%) proposed collaborative multi-user visualization environments. Multi-user systems were deployed in cases where communication played a key role in the educational outcomes of the studies. For example, (Le et al., 2015) applied social virtual reality through the second life platform to enable cooperation, collaboration and social interaction in safety education.

The use of mobile devices for accessing and interacting with visualized contents was considered in 4 articles (7.84%). Two articles made use of immersive CAVE Automatic Virtual Environments (CAVE) (3.92%), projector based displays (3.92%) and 2 articles (3.92%) also implemented immersive Head Mounted Displays (HMDs). All other studies used traditional computer desktops with mouse and keyboard setups to allow users to view and interact with visualization environments (88.24%, 45 articles). 8 studies (15.69%) proposed web based systems and applications. This information is summarized in Figure 4.7.

4.4.4.3 Evaluation layer

The evaluation layer comprised of effectiveness and usability evaluations. Effectiveness evaluations were further categorized as either subjective or objective. In the context of construction education, effectiveness evaluations pertained to learning outcomes, i.e. quantitative improvements in learners' knowledge, skills and performance (e.g. number of errors, or time required to execute an activity), as well as qualitative feedback from subjects. Usability evaluations focused on investigating whether the proposed visualization systems met learners' expectations and requirements. These were implemented through various approaches such as case studies, system trials, use interviews, expert appraisals and so on.

Only 49.07% of the papers analyzed in this study included system evaluations. Regarding the objectiveness of visualization technology evaluation (Figure 4.8), majority of studies on visualization technologies in construction education conduct subjective evaluations (30 articles, 56.60%). Subjective feedback was mostly acquired through interviews, questionnaires

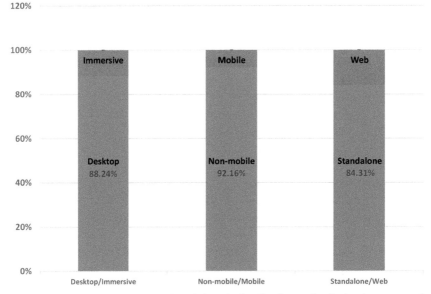

Figure 4.7   Immersive, mobile and web based visualization technologies in construction education.

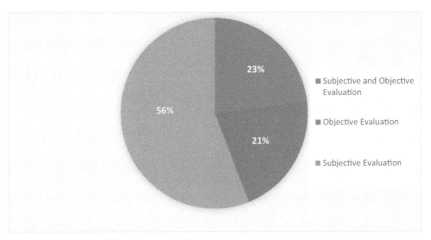

Figure 4.8   Objectiveness in the evaluation of visualization technologies.

and Likert scale ratings. 20.75% of articles provided objective evaluations, while 22.64% (12 articles) combined subjective evaluations with some objective procedures. Objective evaluations applied quantitative approaches such as between subject designs, pre and posttests, performance time

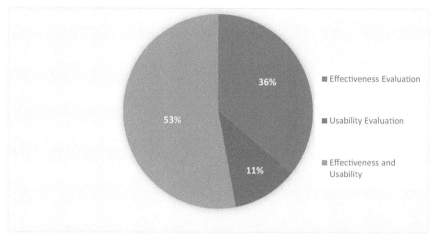

Figure 4.9 Type of evaluation of visualization technologies.

measurements. Results from these evaluations revealed that visualization supported systems have the potential to improve learner motivation, interest, engagement as well as learning outcomes in terms of knowledge acquisition. This suggests there is potential for further success with the application of visualization technologies in construction education. As illustrated in Figure 4.9, 35.84% of the articles analyzed in the evaluation layer conducted effectiveness evaluations (19 articles), while majority of studies (52.83 %, 28 articles) integrated elements of both effectiveness and usability evaluations, and the remaining 11.32% (6 articles) conducted usability evaluations.

4.4.4.4 Adoption layer

The adoption layer concerns the actual implementation and adoption of visualization related systems in construction education programs or courses. Isolated, small-scale case studies which collected subjective feedback from subjects were excluded from this layer. As noted by Wang et al. (2013), real adoption is a critical element in assessing the effectiveness of a tool. This layer involved 21 articles (19.44% of the total number) which significantly involved educational programs, institutions, educators adopting and integrating visualization supported systems. As depicted in Figure 4.10, majority of the proposed systems were implemented as part of short-term projects and courses spanning less than one year (81%). Majority of adoption studies focus on BIM (62%), both as a visualization tool, and as a course with potentials to support visualization in construction education, while the remaining studies

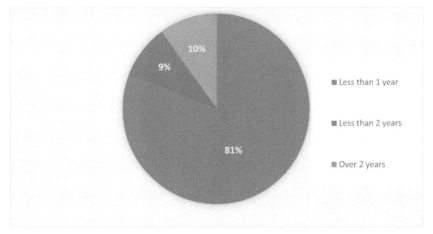

Figure 4.10   Level of adoption of visualization technologies.

incorporated VR-related systems (48%) into curricula. Among the four layers considered, adoption had received the least amount of attention in research. Without the real-world dissemination of visualization technologies in educational institutions, it may remain difficult to obtain feedback which could promulgate large scale adoption.

4.5 Discussion and Summary

Based on the review of 108 articles conducted, numerous discoveries and findings were made. This section rounds up the study with a discussion of main insights, emerging research issues, gaps and potential areas for future research. Firstly, the main insights are summarized as follows:

- Research on visualization technologies in construction education spiked dramatically from 2010 to 2013, and then declined by the end of 2014. Since then research efforts on the field have been consistent, without dramatic increases or decreases.
- Over the years majority of research on the field has been published in conference papers and proceedings, however from 2014, publications have emerged predominantly from journals.
- Regarding prominent application areas, studies have focused on visualization technologies for construction education in general (23.12%), construction management (21.30%), and building construction (14.82%).

- Based on the four layer classification architecture applied, studies were almost evenly distributed in the concept (56 papers), implementation (51 papers) and evaluation layers (53 papers), while adoption received the least attention (21 papers) in literature.
- In terms of concept and theory, most studies proposed new concepts, theories and approaches to apply visualization technologies innovatively (36%). This was followed by development frameworks (30%) and adoption frameworks (25%).
- Analysis of the implementation layer revealed that most developments have been small scale prototypes (52.95%), followed by proof of concepts (39.21%). Full system developments are rare thus far (7.84%). Furthermore, there is very limited research on immersive visualizations (11.76%) and mobile based visualizations (7.84%). Also, few studies have considered Web 3D and web based visualization systems (15.96%).
- Regarding evaluation, most studies (56%) provide subjective evaluations, failing to present quantitative evidence on the impact of visualization technologies. In addition, most studies combined both effectiveness and usability evaluations (53%).
- In the adoption layer, majority of studies with actual implementations in educational programs made use of BIM. This implies that, thus far BIM is the most mature technology, with the highest adoption in construction education.

Secondly, emerging research issues, gaps and potential areas for future research are summarized as follows:

- BIM has been identified as the prominent visualization technology in construction, with numerous powerful commercial tools available on the market. Nowadays, software such as Revit Architecture provide an open API platform for developing specialized programs (Park et al., 2017). Leveraging this, it would be worthwhile for future research efforts to consider BIM-integrated applications with add-in functionalities for specific pedagogic areas in AEC education.
- The development of visualization based systems tends to be complex, costly (Goedert and Rokooei, 2016), and time consuming (Shanbari et al., 2016). Simplifying or automating the process of developing visualized contents for pedagogic purposes could contribute to addressing these issues, however there is a dearth of research efforts in this area. Potential approaches such as laser scanning on real jobsites (Shanbari et al., 2016), or the utilization of BIM and visualization contents from previous construction projects (Pedro et al., 2016) could contribute to

solving this issue. These should be considered in future research in order to address the prohibitive costs and efforts required for content development.

- Research efforts in construction education have not adequately considered interoperability between visualization based systems, in order to provide learners with a wider range of pedagogic contents. (Hilfert and König, 2016) recently touched on this issue, and attempted to create a unified, multipurpose environment for construction and engineering through BIM and VR. Leveraging the wide data exchange capabilities of the Industry Foundation Classes (IFC) in BIM, the approach enabledmetadata enhanced virtual environments, to offer users more profound and interactive experiences. Enabling this type of inter-operability in construction education could contribute greatly to enrich learner experiences.

- Limited reference materials, component databases and librariesare some of the challenges encountered with deploying BIM in construc-tion education (Wong et al., 2011). Adopting object-oriented database approaches for modelling in future developments could afford enhanced collaborative environments (Abrishami et al., 2013). Research not only on content databases and libraries, but also automated recommenda-tion systems based on ontologies and semantic web technologies could contribute greatly to improving access to educational contents.

- Immersive visualizations for construction education have not been considered in depth thus far. Recently, VR-glasses have emerged, pre-senting immense potential to transform educational experiences and afford new opportunities for improving learning outcomes in tertiary education. Now that these are more affordable and accessible than ever, their potential and impact on construction education should be considered.

- There is a deficit of research which thoroughly assess the usability of visualization technologies for construction education. This point is corroborated by the limited research efforts on evaluation frameworks in the field. More research efforts on this topic could contribute to the development of more robust and intuitive systems in the future.

- Few studies consider the long term effects of visualization technologies on learning outcomes and knowledge retention. Such evidence would be useful in demonstrating the impacts of visualization technologies. Objective evaluation results could help motivate the larger scale adop-tion and implementation of visualization technologies within existing curricula.

- Systematic and carefully designed approaches are crucial to ensure new technologies are embedded and integrated strategically into curricula (Horne and Thompson, 2008). However, there is a lack of literature and research findings on the factors affecting buy-in, and institutional support for integrating visualization technologies into programs and curricula. Further studies and documentation on these factors would be worthwhile to shed light on nature of visualization technology adoption in tertiary construction curricula.

4.6 Conclusion

Construction personnel constantly face unexpected challenges in meeting the demands of society. Hence, it is necessary that construction graduates are equipped with the knowledge and skills required to meet the needs of construction work nowadays. The use of advanced visualization technologies is rapidly growing in entertainment and education, and permeating into industry practice in various disciplines. Over recent years, construction industry and education-oriented research alike have considered applications of visualization technologies for a variety of goals. With the rapid advance in these technologies, it is possible to invigorate construction education and provide access to high impact contents in unprecedented innovative ways. Considerable efforts have focused on innovatively applying visualization techniques to address the limitations of conventional construction education. This paper comprehensively synthesizes research efforts on visualization technologies in construction education with a systematic review of studies from 2005 to 2016. Through a systematic search strategy, 108 papers were found based on defined eligibility criteria. In order to summarize past research works, identify issues, gaps, and areas of concentration, a four layer architecture which classified studies based on the concept and theory, development, evaluation and adoption layers was applied. The findings of this review provide insights for educators and researchers into: the different ways visualization technologies have been adopted innovatively; the different types of systems that have been implemented; and the level of evaluation and extent of adoption in actual educational institutions. The comprehensive review and analyses presented in this paper serve as a foundation to pave the way for future research on advanced visualization technologies in construction education. The authors believe that further research efforts focusing on developing and integrating visualization technologies with construction education could contribute greatly in preparing graduates to meet and fulfil the demands of the construction industry.

References

R. Sacks and E. Pikas, "Building Information Modeling Education for Construction Engineering and Management. I: Industry Requirements, State of the Art, and Gap Analysis," *J. Constr. Eng. Manag.*, vol. 139, no. 11, p. 04013016, Nov. 2013.

R. Hussain, A. Pedro, D. Y. Lee, H. C. Pham, and C. S. Park, "Impact of safety training and interventions on training-transfer: targeting migrant construction workers," *Int. J. Occup. Saf. Ergon.*, pp. 1–13, June 2018.

J. Teizer, B. S. Allread, C. E. Fullerton, and J. Hinze, "Autonomous pro-active real-time construction worker and equipment operator proximity safety alert system," *Autom. Constr.*, vol. 19, no. 5, pp. 630–640, Aug. 2010.

R. Hussain, D. Y. Lee, H. C. Pham, and C. S. Park, "Safety regulation classification system to support BIM based safety management," in *ISARC 2017 – Proceedings of the 34th International Symposium on Automation and Robotics in Construction*, 2017, no. Isarc, pp. 30–36.

S. Zhang, J. Teizer, J.-K. Lee, C. M. Eastman, and M. Venugopal, "Building Information Modeling (BIM) and Safety: Automatic Safety Checking of Construction Models and Schedules," *Autom. Constr.*, vol. 29, pp. 183–195, Jan. 2013.

A. Pedro, Q. T. Le, and C. S. Park, "Framework for Integrating Safety into Construction Methods Education through Interactive Virtual Reality," *J. Prof. Issues Eng. Educ. Pract.*, vol. 142, no. 2, p. 04015011, Apr. 2016.

H. Pham, N.-N. Dao, J.-U. Kim, S. Cho, and C.-S. Park, "Energy-Efficient Learning System Using Web-Based Panoramic Virtual Photoreality for Interactive Construction Safety Education," *Sustainability*, vol. 10, no. 7, p. 2262, July 2018.

C. S. Park, Q. T. Le, A. Pedro, and C. R. Lim, "Interactive Building Anatomy Modeling for Experiential Building Construction Education," *J. Prof. Issues Eng. Educ. Pract.*, vol. 142, no. 3, p. 04015019, July 2016.

D. Bouchlaghem, H. Shang, J. Whyte, and A. Ganah, "Visualisation in architecture, engineering and construction (AEC)," *Autom. Constr.*, vol. 14, no. 3, pp. 287–295, June 2005.

W. Zhou, J. Whyte, and R. Sacks, "Construction safety and digital design: A review," *Autom. Constr.*, vol. 22, pp. 102–111, Mar. 2012.

S. Bhoir and B. Esmaeili, "State-of-the-Art Review of Virtual Reality Environment Applications in Construction Safety," in *AEI*, 2015, pp. 457–468.

A. A. Deshpande and S. H. Huang, "Simulation games in engineering education: A state-of-the-art review," *Comput. Appl. Eng. Educ.*, vol. 19, no. 3, pp. 399–410, Sep. 2011.

H.-L. Chi, S.-C. Khang, and X. Wang, "Research trends and opportunities of augmented reality applications in architecture, engineering, and construction," *Autom. Constr.*, vol. 33, pp. 116–122, Aug. 2013.

S. Rankohi and L. Waugh, "Review and analysis of augmented reality literature for construction industry," *Vis. Eng.*, vol. 1, no. 1, p. 9, Aug. 2013.

X. Wang, M. J. Kim, P. E. D. Love, and S.-C. Kang, "Augmented Reality in built environment: Classification and implications for future research," *Autom. Constr.*, vol. 32, pp. 1–13, July 2013.

G. Keenaghan and I. Horváth, "State of the Art of Using Virtual Reality Technologies in Built Environment Education," in *Proceedings of TMCE*, 2014, pp. 935–948.

Q. T. Le, A. Pedro, and C. S. Park, "A Social Virtual Reality Based Construction Safety Education System for Experiential Learning," *J. Intell. Robot. Syst.*, vol. 79, no. 3–4, pp. 487–506, Aug. 2015.

A. Pedro, H. C. Pham, J. U. Kim, and C. Park, "Development and Evaluation of Context-based Assessment System for Visualization-Enhanced Construction Safety Education," *Int. J. Occup. Saf. Ergon.*, pp. 1–33, Nov. 2018.

A. Shirazi and A. H. Behzadan, "Content delivery using augmented reality to enhance students' performance in a building design and assembly project," *Adv. Eng. Educ.*, vol. 4, no. 3, pp. 1–24, 2015.

J. Irizarry, P. Meadati, W. S. Barham, and A. Akhnoukh, "Exploring Applications of Building Information Modeling for Enhancing Visualization and Information Access in Engineering and Construction Education Environments," *Int. J. Constr. Educ. Res.*, vol. 8, no. 2, pp. 119–145, Apr. 2012.

P. Forsythe, "The Construction Game – Using Physical Model Making to Simulate Realism in Construction Education," *J. Educ. Built Environ.*, vol. 4, no. 1, pp. 57–74, July 2009.

D. Nikolic, S. Jaruhar, and J. I. Messner, "Educational Simulation in Construction: Virtual Construction Simulator," *J. Comput. Civ. Eng.*, vol. 25, no. 6, pp. 421–429, Nov. 2011.

B. Becerik-Gerber, A.M.ASCE, K. Ku, and F. Jazizadeh, "BIM-Enabled Virtual and Collaborative Construction Engineering and Management," *J. Prof. Issues Eng. Educ. Pract.*, vol. 138, no. 3, pp. 234–245, July 2012.

Y.-L. Liaw, K.-Y. Lin, M. Li, and N.-W. Chi, "Learning Assessment Strategies for an Educational Construction Safety Video Game," in *Construction Research Congress*, 2012, pp. 2091–2100.

H. Shanbari, N. Blinn, and R. R. A. Issa, "Using augmented reality video in enhancing masonry and roof component comprehension for construction management students," *Eng. Constr. Archit. Manag.*, vol. 23, no. 6, pp. 765–781, Nov. 2016.

V. Vimonsatit and T. Htut, "Civil Engineering students' response to visualisation learning experience with building information model," *Australas. J. Eng. Educ.*, vol. 21, no. 1, pp. 27–38, Jan. 2016.

J. L. Perdomo, M. F. Shiratuddin, W. Thabet, and A. Ananth, "Interactive 3D Visualization As A Tool For Construction Education," in *2005 6th International Conference on Information Technology Based Higher Education and Training*, pp. FB4-23–FB4-28.

K. D. A. Wong, K. W. F. Wong, and A. Nadeem, "Building information modelling for tertiary construction education in Hong Kong," *Electron. J. Inf. Technol. Constr.*, vol. 16, pp. 467–476, 2011.

M. E. Haque, D. Ph, and P. E. M. E. Haque, "4 D Construction Visualization: Techniques With Examples," in *2007 Annual Conference & Exposition*, 2007.

J. Goedert, Y. Cho, M. Subramaniam, H. Guo, and L. Xiao, "A framework for Virtual Interactive Construction Education (VICE)," *Autom. Constr.*, vol. 20, no. 1, pp. 76–87, Jan. 2011.

H. C. Pham et al., "Virtual field trip for mobile construction safety education using 360-degree panoramic virtual reality," *Int. J. Eng. Educ.*, vol. 34, no. 4, pp. 1174–1191, 2018.

H. C. Pham, A. Pedro, Q. T. Le, D.-Y. Lee, and C.-S. Park, "Interactive safety education using building anatomy modelling," *Univers. Access Inf. Soc.*, pp. 1–17, Nov. 2017.

A. Pedro, P. H. Chien, and C. S. Park, "Towards a Competency-based Vision for Construction Safety Education," *IOP Conf. Ser. Earth Environ. Sci.*, vol. 143, no. 1, p. 012051, Apr. 2018.

A. Z. Sampaio, P. G. Henriques, and C. O. Cruz, "Interactive models used in Civil Engineering education based on virtual reality technology," in *2nd Conference on Human System Interactions*, 2009, pp. 170–176.

C. Clevenger, S. Glick, and C. L. del Puerto, "Interoperable Learning Leveraging Building Information Modeling (BIM) in Construction Education," *Int. J. Constr. Educ. Res.*, vol. 8, no. 2, pp. 101–118, Apr. 2012.

J. C. Aurich, D. Ostermayer, and C. H. Wagenknecht, "Improvement of manufacturing processes with virtual reality-based CIP workshops," *Int. J. Prod. Res.*, vol. 47, no. 19, pp. 5297–5309, Oct. 2009.

A. Z. Sampaio and O. P. Martins, "The application of virtual reality technology in the construction of bridge: The cantilever and incremental launching methods," *Autom. Constr.*, vol. 37, pp. 58–67, Jan. 2014.

K. Y. Lin, J. W. Son, and E. M. Rojas, "A pilot study of a 3D game environment for construction safety education," *Electron. J. Inf. Technol. Constr.*, vol. 16, no. March 2010, pp. 69–83, 2011.

H. Guo, H. Li, G. Chan, and M. Skitmore, "Using game technologies to improve the safety of construction plant operations," *Accid. Anal. Prev.*, vol. 48, pp. 204–213, Sep. 2012.

L. Rebenitsch and C. Owen, "Review on cybersickness in applications and visual displays," *Virtual Real.*, vol. 20, no. 2, pp. 101–125 June 2016.

C.-S. Park and H.-J. Kim, "A framework for construction safety management and visualization system," *Autom. Constr.*, vol. 33, pp. 95–103, Aug. 2013.

O.-S. Kwon, C.-S. Park, and C.-R. Lim, "A defect management system for reinforced concrete work utilizing BIM, image-matching and augmented reality," *Autom. Constr.*, vol. 46, pp. 74–81, Oct. 2014.

A. H. Behzadan and V. R. Kamat, "Enabling discovery−based learning in construction using telepresent augmented reality," *Autom. Constr.*, vol. 33, pp. 3–10, Aug. 2013.

S. K. Ayer, J. I. Messner, and C. J. Anumba, "Augmented Reality Gaming in Sustainable Design Education," *J. Archit. Eng.*, vol. 22, no. 1, p. 04015012, Mar. 2016.

K. Ku and P. S. Mahabaleshwarkar, "Building interactive modeling for construction education in virtual worlds," *Electron. J. Inf. Technol. Constr.*, vol. 16, no. September 2010, pp. 189–208, 2011.

M. J. Grant and A. Booth, "A typology of reviews: an analysis of 14 review types and associated methodologies," *Heal. Inf. Libr. J.*, vol. 26, no. 2, pp. 91–108, June 2009.

A. A. Chadegani *et al.*, "A Comparison between Two Main Academic Literature Collections: Web of Science and Scopus Databases," *Asian Soc. Sci.*, vol. 9, no. 5, p. 18, Apr. 2013.

P. Jacso, "The pros and cons of computing the h−index using Google Scholar," *Online Inf. Rev.*, vol. 32, no. 3, pp. 437–452, June 2008.

B. Kitchenham, "Procedures for Performing Systematic Reviews," 2004.

L. Duckworth, T. Sulbaran, and A. P. Strelzof, "Usability of a Collaborative Virtual Reality Environment Earthwork Exercises," in *2012 ASEE Annual Conference & Exposition*, 2012.

D. Nikolic, S. Jaruhar, and J. I. Messner, "An Educational Simulation in Construction: The Virtual Construction Simulator," in *Computing in Civil Engineering*, 2009, pp. 633–642.

E. A. Obonyo, "An agent–based intelligent virtual learning environment for construction management," *Constr. Innov.*, vol. 11, no. 2, pp. 142–160, Apr. 2011.

C.-S. Park, H.-J. Kim, H.-T. Park, J.-H. Goh, and A. Pedro, "BIM-based idea bank for managing value engineering ideas," *Int. J. Proj. Manag.*, vol. 35, no. 4, pp. 699–713, May 2017.

J. D. Goedert and S. Rokooei, "Project-Based Construction Education with Simulations in a Gaming Environment," *Int. J. Constr. Educ. Res.*, vol. 12, no. 3, pp. 208–223, July 2016.

H. A. Shanbari, N. M. Blinn, and R. R. Issa, "Laser scanning technology and BIM in construction management education," *J. Inf. Technol. Constr.*, vol. 21, no. November 2015, pp. 204–217, 2016.

T. Hilfert and M. König, "Low-cost virtual reality environment for engineering and construction," *Vis. Eng.*, vol. 4, no. 1, p. 2, Dec. 2016.

S. Abrishami, J. S. Goulding, A. Ganah, and F. P. Rahimian, "Exploiting Modern Opportunities in AEC Industry: A Paradigm of Future Opportunities," in *AEI*, 2013, pp. 321–333.

M. Horne and E. M. Thompson, "The Role of Virtual Reality in Built Environment Education," *J. Educ. Built Environ.*, vol. 3, no. 1, pp. 5–24, July 2008.

A. Z. Sampaio and P. G. Henriques, "Visual simulation of civil engineering activities: Didactic virtual models," in *16th International Conference in Central Europe on Computer Graphics, Visualization and Computer Vision, WSCG'2008 – In Co-operation with EUROGRAPHICS, Full Papers*, 2008, pp. 143–149.

Bibliography

J. Irizarry, M. Gheisari, S. Zolfagharian, and P. Meadati, "Human Computer Interaction Modes for Construction Education Applications: Experimenting with Small Format Interactive Displays," *Int. J. Constr. Educ. Res.*, vol. 9, no. 2, pp. 83–101, Apr. 2013.

S. Azhar, A. Behringer, A. Sattineni, and T. Maqsood, "BIM for facilitating construction safety planning and management at jobsites," in *Proceedings of the CIB-W099 International Conference: Modelling and 620 Building Safety*, 2012, pp. 82–92.

M. J. Kim, X. Wang, P. E. D. Love, H. Li, and S. C. Kang, "Virtual reality for the built environment: A critical review of recent advances," *J. Inf. Technol. Constr.*, vol. 18, no. August, pp. 279–305, 2013.

J. Hong, E. Suh, and S.-J. Kim, "Context-aware systems: A literature review and classification," *Expert Syst. Appl.*, vol. 36, no. 4, pp. 8509–8522, May 2009.

Sampaio Alcínia Z., P. G. Henriques, and P. S. Ferreira, "Virtual Reality Models Used in Civil Engineering Education," in *Proceedings of the 24th IASTED International Conference on Internet and Multimedia Systems and Applications*, 2006, pp. 119–124.

C. Clevenger, M. E. Ozbek, S. Glick, and D. Porter, "Integrating BIM into Construction Management Education Integrating BIM into Construction Management Education," in *EcoBuild Proceedings of the BIM-Related Academic Workshop*, 2010, no. June 2016, pp. 1–8.

M. Haque and R. Moosa, "AC 2009-419: Virtual Walk Through of a Building Foundation System Using Game Engine," in *Annual Conference & Exposition*, 2009.

M. E. Haque and R. Chanda, "Desk-Top Virtual Reality for Teaching of Steel Structures," in *Internet and Multimedia Systems, and Applications*, 2005.

W. Zhou, D. Heesom, P. Georgakis, C. Nwagboso, and A. Feng, "An Interactive Approach To Collaborative 4D Construction Planning," *J. Inf. Technol. Constr.*, vol. 14, no. March, pp. 30–47, 2009.

S. Zolfagharian, M. Gheisari, J. Irizarry, and P. Meadati, "Exploring the impact of various interactive displays on student learning in construction courses," in *ASEE Annual Conference and Exposition*, 2013.

D. Zhao, K. Sands, Z. Wang, and Y. Ye, "Building information modeling-enhanced team-based learning in construction education," in *2013 12th International Conference on Information Technology Based Higher Education and Training (ITHET)*, 2013, pp. 1–5.

T. Sulbaran and L. F. Jones III, "Utilizing a collaborative virtual reality environment as a training tool for construction students," in *ASEE Annual Conference and Exposition, Conference Proceedings*, 2012.

M. F. Shiratuddin, "Integrating computer game-based learning into construction education," in *2011 International Conference on Information Technology and Multimedia: "Ubiquitous ICT for Sustainable and Green Living", ICIM*, 2011, no. November 2011.

G. Wang, H. Liu, and S. Zhang, "Research on a framework for a BIM-based construction engineering and project management education platform," *World Trans. Eng. Technol. Educ.*, vol. 12, no. 1, pp. 101–104, 2014.

A. J. P. Tixier, A. Albert, and M. R. Hallowell, "Teaching construction hazard recognition through high fidelity augmented reality," in *ASEE Annual Conference and Exposition, Conference Proceedings*, 2013.

W. Wu and Y. Luo, "Pedagogy and assessment of student learning in BIM and sustainable design and construction," *J. Inf. Technol. Constr.*, vol. 21, no. November 2015, pp. 218–232, 2016.

R. Zhang, "Design and Implementation of Construction Engineering Teaching System Based Virtual Reality," *Appl. Mech. Mater.*, vol. 353–356, pp. 3634–3639, Aug. 2013.

Y.-C. Wu and J.-C. Yang, "The Effects of the Social Media-Based Activities and Gaming-Based Learning in Construction Education," in *2015 IIAI 4th International Congress on Advanced Applied Informatics*, 2015, pp. 323–328.

W. Wu and I. Kaushik, "A BIM-based educational gaming prototype for undergraduate research and education in design for sustainable aging," in *2015 Winter Simulation Conference (WSC)*, 2015, pp. 1091–1102.

A. Strelzoff and T. Sulbaran, "Transformation of a Collaborative Virtual Reality Environment for construction scheduling to help individual with mental health issues," in *2008 38th Annual Frontiers in Education Conference*, 2008, p. S4J–13–S4J–18.

J. Son, K.-Y. Lin, and E. M. Rojas, "Developing and Testing a 3D Video Game for Construction Safety Education," in *Computing in Civil Engineering*, 2011, pp. 867–874.

A. Shirazi and A. H. Behzadan, "Design and Assessment of a Mobile Augmented Reality-Based Information Delivery Tool for Construction and Civil Engineering Curriculum," *J. Prof. Issues Eng. Educ. Pract.*, vol. 141, no. 3, p. 04014012, July 2015.

A. Shirazi and A. H. Behzadan, "Technology-enhanced learning in construction education using mobile context-aware augmented reality visual simulation," in *2013 Winter Simulations Conference (WSC)*, 2013, pp. 3074–3085.

A. Z. Sampaio and P. G. Henriques, "Simulation of Construction Processes Within Virtual Environments," in *Proceedings of the Third International Conference on Computer Graphics Theory and Applications*, 2008, pp. 326–331.

A. Z. Sampaio, A. N. A. R. Gomes, J. P. Santos, and D. P. Rosa, "Virtual Reality Applied on Civil Engineering Education: Construction Activity Supported on Interactive Models," *Int. J. Eng. Educ.*, vol. 29, no. 6, pp. 1331–1347, 2013.

A. Z. Sampaio and L. Viana, "Virtual Reality used as a learning technology: Visual simulation of the construction of a bridge deck," in *Information Systems and Technologies (CISTI)*, 2013, pp. 1–5.

A. Z. Sampaio, P. G. Henriques, and C. O. Cruz, "Virtual Didactic Models on engineering education: Construction of a wall, a bridge and a roof," in *Proceedings of the IADIS International Conference Computer Graphics, Visualization, Computer Vision and Image Processing 2009, CGVCVIP 2009. Part of the IADIS MCCSIS 2009*, 2009.

M. Setareh, D. A. Bowman, A. Kalita, M. Gracey, and J. Lucas, "Application of a Virtual Environment System in Building Sciences Education," *J. Archit. Eng.*, vol. 11, no. 4, pp. 165–172, Dec. 2005.

A. Z. Sampaio, P. Henriques, and P. Studer, "Learning construction processes using virtual reality models," *Electron. J. Inf. Technol. Constr.*, vol. 10, no. October 2004, pp. 141–151, 2005.

A. Z. Sampaio, M. M. Ferreira, D. P. Rosário, and O. P. Martins, "3D and VR models in Civil Engineering education: Construction, rehabilitation and maintenance," *Autom. Constr.*, vol. 19, no. 7, pp. 819–828, Nov. 2010.

R. Sacks and R. Barak, "Teaching Building Information Modeling as an Integral Part of Freshman Year Civil Engineering Education," *J. Prof. Issues Eng. Educ. Pract.*, vol. 136, no. 1, pp. 30–38, Jan. 2010.

E. Redondo, I. Navarro, A. Sánchez, and D. Fonseca, "Visual Interfaces and User Experience: Augmented Reality for Architectural Education: One Study Case and Work in Progress," Springer, Berlin, Heidelberg, 2011, pp. 355–367.

E. Pikas, R. Sacks, and O. Hazzan, "Building Information Modeling Education for Construction Engineering and Management. II: Procedures and Implementation Case Study," *J. Constr. Eng. Manag.*, vol. 139, no. 11, p. 05013002, Nov. 2013.

F. Peterson, T. Hartmann, R. Fruchter, and M. Fischer, "Teaching construction project management with BIM support: Experience and lessons learned," *Autom. Constr.*, vol. 20, no. 2, pp. 115–125, Mar. 2011.

R. Palomera-Arias, "Building information modeling laboratory exercises in a construction science and management building systems course," in *2015 IEEE Frontiers in Education Conference (FIE)*, 2015, pp. 1–6.

A. Nejat, M. M. Darwish, and T. Ghebrab, "BIM Teaching Strategy for Construction Engineering Students," in *2012 ASEE Annual Conference & Exposition*, 2012.

S. L. J. I. M. C. A. Dragana Nikolic, "The Virtual Construction Simulator: Evaluating an Educational Simulation Application for Teaching

Construction Management Concepts," *CIB W78 2010 - Appl. IT AEC Ind.*, 2010.

N. O. Nawari, "The Role of BIM in Teaching Structural Design," in *Structures Congress*, 2015, pp. 2622–2631.

I. Mutis and R. R. A. Issa, "Enhancing Spatial and Temporal Cognitive Ability in Construction Education Through Augmented Reality and Artificial Visualizations," in *Computing in Civil and Building Engineering (2014)*, 2014, pp. 2079–2086.

M. Maghiar, S. Jain, and J. G. Sullivan, "Strategy to incorporate BIM curriculum in planning and scheduling classes," in *120th ASEE Annual Conference & Exposition*, 2013.

S. Moaven and K. C. Chou, "An interactive steel connection teaching tool – A virtual structure," in *2014 ASEE Annual Conference & Exposition*, 2014.

J. I. Messner, D. R. Riley, and M. J. Horman, "An Interactive Visualization Environment for Construction Engineering Education," in *Construction Research Congress*, 2005, pp. 1–10.

O. Martins and A. Z. Sampaio, "The Incremental Launching Method for Educational Virtual Model," Springer, Berlin, Heidelberg, 2009, pp. 329–332.

W. Lu, Y. Peng, Q. Shen, and H. Li, "Generic Model for Measuring Benefits of BIM as a Learning Tool in Construction Tasks," *J. Constr. Eng. Manag.*, vol. 139, no. 2, pp. 195–203, Feb. 2013.

C. Livingston, "From CAD to BIM: Constructing Opportunities in Architectural Education," in *AEI*, 2008, pp. 1–9.

N. Lee and C. Dossick, "Leveraging Building Information Modeling Technology in Construction Engineering and Management Education," in *2012 ASEE Annual Conference & Exposition*, 2012.

N. Lee and D. Hollar, "Probing BIM Education in Construction Engineering and Management Programs Using Industry Perceptions," in *49th ASC Annual International Conference Proceedings*, 2013, pp. 1–8.

S. Lee, D. Nikolic, J. I. Messner, and C. J. Anumba, "The Development of the Virtual Construction Simulator 3: An Interactive Simulation Environment for Construction Management Education," in *Computing in Civil Engineering (2011)*, 2011, pp. 454–461.

N. Lee, C. S. Dossick, and S. P. Foley, "Guideline for Building Information Modeling in Construction Engineering and Management Education," *J. Prof. Issues Eng. Educ. Pract.*, vol. 139, no. 4, pp. 266–274, Oct. 2013.

Q. T. Le, A. Pedro, H. C. Pham, and C. S. Park, "A Virtual World Based Construction Defect Game for Interactive and Experiential Learning," *Int. J. Eng. Educ.*, vol. 32, no. 1, pp. 457–467, 2016.

Q.-T. Le and Chan-Sik Park, "Construction safety education model based on second life," in *Proceedings of IEEE International Conference on Teaching, Assessment, and Learning for Engineering (TALE) 2012*, 2012, p. H2C–1–H2C–5.

K. Ku and Y. Gaikwat, "Construction Education in Second Life," in *Construction Research Congress 2009*, 2009, pp. 1378–1387.

J. R. Juang, W. H. Hung, and S. C. Kang, "Using game engines for physics-based simulations – A forklift," *Electron. J. Inf. Technol. Constr.*, vol. 16, no. January 2011, pp. 3–22, 2011.

S. Karshenas and D. Haber, "Developing a Serious Game for Construction Planning and Scheduling Education," in *Construction Research Congress*, 2012, no. May, pp. 2042–2051.

H. Kim, "Energy modeling / Simulation Using the BIM technology in the Curriculum of Architectural and Construction Engineering and Management," in *ASEE Annual Conference and Exposition, Conference Proceedings*, 2013.

D. Tranfield, "Procedures for performing systematic reviews," *Br. J. Manag.*, vol. 14, no. 0, pp. 207–222, 2003.

J.-L. L. Kim, "Effectiveness of green-BIM teaching method in construction education curriculum," in *2014 ASEE Annual Conference & Exposition*, 2014.

J.-L. Kim, "Use of BIM for Effective Visualization Teaching Approach in Construction Education," *J. Prof. Issues Eng. Educ. Pract.*, vol. 138, no. 3, pp. 214–223, July 2012.

P. Jacsó, "Google Scholar revisited," *Online Inf. Rev.*, vol. 32, no. 1, pp. 102–114, Feb. 2008.

M. Horne and N. Hamza, "Integration of virtual reality within the built environment curriculum," *Electron. J. Inf. Technol. Constr.*, vol. 11, pp. 311–324, 2006.

A. Ghosh, "Virtual Construction + Collaboration Lab: Setting a New Paradigm for BIM Education," in *2012 ASEE Annual Conference & Exposition*, 2012.

L. F. Gül, N. Gu, and A. Williams, "Virtual worlds as a constructivist learning platform: Evaluations of 3D virtual worlds on design teaching and learning," *Electron. J. Inf. Technol. Constr.*, vol. 13, no. June, pp. 578–593, 2008.

A. Ghosh, K. Parrish, and A. D. Chasey, "From BIM to collaboration: A proposed integrated construction curriculum," *ASEE Annu. Conf. Expo. June 23, 2013 – June 26, 2013*, 2013.

M. Haque, "Multi-dimensional Construction Visualizations with Examples: Suggested Topics for Graduate Course," in *American Society for Engineering Education*, 2010, pp. 401–408.

J. D. Goedert, R. Pawloski, S. Rokooeisadabad, and M. Subramaniam, "Project-Oriented Pedagogical Model for Construction Engineering Education Using Cyberinfrastructure Tools," *J. Prof. Issues Eng. Educ. Pract.*, vol. 139, no. 4, pp. 301–309, Oct. 2013.

S. Glick, D. Porter, and C. Smith, "Student Visualization: Using 3-D Models in Undergraduate Construction Management Education," *Int. J. Constr. Educ. Res.*, vol. 8, no. 1, pp. 26–46, Jan. 2012.

P. Forsythe, J. Jupp, and A. Sawhney, "Building Information Modelling in Tertiary Construction Project Management Education: A Programme-wide Implementation Strategy," *J. Educ. Built Environ.*, vol. 8, no. 1, pp. 16–34, Dec. 2013.

C. S. Dossick, N. Lee, and S. Foleyk, "Building Information Modeling in Graduate Construction Engineering and Management Education," in *Computing in Civil and Building Engineering (2014)*, 2014, pp. 2176–2183.

S. Dong, A. H. Behzadan, F. Chen, and V. R. Kamat, "Collaborative visualization of engineering processes using tabletop augmented reality," *Adv. Eng. Softw.*, vol. 55, pp. 45–55, Jan. 2013.

H. Dib and N. Adamo-Villani, "Serious Sustainability Challenge Game to Promote Teaching and Learning of Building Sustainability," *J. Comput. Civ. Eng.*, vol. 28, no. 5, p. A4014007, Sep. 2014.

C. Clevenger, C. L. Del Puerto, and S. Glick, "Interactive BIM-enabled safety training piloted in construction education," *Adv. Eng. Educ.*, vol. 4, no. 3, pp. 1–14, 2015.

A. Chegu Badrinath, Y. Chang, and S. Hsieh, "A review of tertiary BIM education for advanced engineering communication with visualization," *Vis. Eng.*, vol. 4, no. 1, p. 9, Dec. 2016.

J. Bozoglu, "Collaboration and coordination learning modules for BIM education," *J. Inf. Technol. Constr.*, vol. 21, no. July, pp. 152–163, 2016.

J. Boon and C. Prigg, "Releasing the potential of BIM in construction education," *CIB, Manag. Innov. a Sustain. Built Environ.*, no. 20–23 June, 2011.

A. Chasey, H. Ciszczon, A. Ghosh, and L. Hogle, "Evolution of the new construction classroom," in *IMETI 2012 – 5th International Multi-Conference on Engineering and Technological Innovation, Proceedings*, 2012, no. 2012, pp. 19–23.

F. Castronovo, D. Nikolic, S. E. Zappe, R. M. Leicht, and J. I. Messner, "Enhancement of Learning Objectives in Construction Engineering Education: A Step toward Simulation Assessment," in *Construction Research Congress*, 2014, pp. 339–348.

S. Berwald, "From CAD to BIM: The Experience of Architectural Education with Building Information Modeling," in *AEI*, 2008, pp. 1–5.

A. H. Behzadan, A. Iqbal, and V. R. Kamat, "A collaborative augmented reality based modeling environment for construction engineering and management education," in *Proceedings of the 2011 Winter Simulation Conference (WSC)*, 2011, pp. 3568–3576.

F. Manzoor Arain, M. Burkle, and C. Chair, "Learning construction project management in the virtual world: leveraging on second life," *J. Inf. Technol. Constr.*, vol. 16, pp. 243–258, 2011.

H. Alshanbari and R. R. A. Issa, "Use of Video Games to Enhance Construction Management Education," in *Computing in Civil and Building Engineering*, 2014, pp. 2135–2142.

K. Alanne, "An overview of game-based learning in building services engineering education," *Eur. J. Eng. Educ.*, vol. 41, no. 2, pp. 204–219, Mar. 2016.

W. A. Abdelhameed, "Virtual Reality Use in Architectural Design Studios: A Case of Studying Structure and Construction," *Procedia Comput. Sci.*, vol. 25, pp. 220–230, Jan. 2013.

5

A Legal Framework and Compliance with Construction Safety Laws and Regulations in Vietnam

Thi-Thanh-Mai Pham[1], Quang-Vu Pham[2], Anh-Tuan Pham-Hang[3] and Hai Chien Pham[4*]

[1]Faculty of International Trade, College of Foreign Economic Relation, Ho Chi Minh City 7000000, Vietnam
[2]Faculty of Construction, Ho Chi Minh City of Transport and Communications, Ho Chi Minh City 7000000, Vietnam
[3]School of Computer Science and Engineering, International University – Vietnam National University HCMC, Ho Chi Minh City 7000000, Vietnam
[4]Faculty of Civil Engineering, Ton Duc Thang University, Ho Chi Minh City 7000000, Vietnam
E-mail: phammai.tmqt@gmail.com; vu.phamquang258@gmail.com; anhtuanphamhang@gmail.com; phamhaichien@tdtu.edu.vn
*Corresponding Author

Construction in Vietnam is acknowledged as one of the most hazardous and dangerous industries, which accounts for very high rate of accidents and injuries. Despite given great attention by Government recently, the number of construction accidents is around 30% of occupational accidents of the whole industries. To fully understand current construction in Vietnam, this chapter therefore focuses on investigating Vietnamese construction safety for 5 years from 2013 to 2017. Through investigation, this chapter reveals that construction safety laws and regulations vary and are promulgated by not only Vietnam Ministry of Construction (MOC) but also other Government ministries and agencies. Thus, construction firms are embarrassed when complying with safety legal for assuring safety at jobsite. With this regard, the chapter aims to develop Construction Safety Legal Framework (CSLF), providing a comprehensive viewpoint of construction safety legislation hierarchy linked with state governance hierarchy of Vietnam. After that, an investigation into the compliance with construction safety laws and regulations is conducted in order to analyse key compliance and non-compliance in practice. Through this, recommendations are discussed in order to promote construction safety in Vietnam.

5.1 Introduction

A global forecast for the construction industry to 2025 reports that construction plays a significant role in economic development of nations over the world, contributing $8.7 trillion and accounting for 12.2% of the world's economic (Perspectives and Economics 2013). Despite the important economic contribution to Gross Domestic Product (GDP) (OSHA, 2018), construction sector is associated with very high amount of accidents due to its complex and dangerous nature (Zaira and Hadikusumo, 2017), (Pham, 2018a), (Gambatese, 2008). According to a report of the US Bureau of Labor Statistics in 2016, there are approximately 900 fatal and 200,000 non-fatal injuries in construction industry (Bureau of Labor Statistics, 2018). Seriously, a total of 60,000 fatal construction injuries happened per year worldwide, which interprets to one injury in every nine minutes (Lingard et al., 2013). In fact, construction industry causes 31% of all fatalities worldwide (Reese, 2006). The high rate of construction accidents and fatal injuries causes many problems related to cost overruns and time delays (Pham, 2017), (Pedro, 2015), which negatively affect project performance (Hussain, 2018), (Le, 2016). Despite the attention paid to construction jobsite, the number of accidents in construction has been reported to be twice the industrial average (Le, 2014; Rowlinson, 2004). Therefore, construction safety has been a concern to not only practitioners but also researchers in order to improve safety work performance at the jobsite.

As a developing country, Vietnam government therefore has given much attention to promote construction to the top three leading industries, accounting for approximately 10% of GDP in recent years (MOC, 2018). However, Vietnam is facing a high number of construction accidents and fatalities. According to annual reports of the Ministry of Labour, War invalids and Social Affairs (MOLISA), construction sector has the highest rate of accidents, accounting for approximately 20.8% of total accidents in Vietnam (MOLISA, 2018). The MOLISA reports that root causes of construction accidents are a lack of comprehensively enforcing safety legal documents issued by Vietnam government as well as safety guides of project. Although Vietnam Ministry of Construction (MOC) is a key ministry accountable for state management on all construction aspects including safety, construction safety laws and regulations vary and promulgate by not only MOC but also other Government ministries and agencies such as MOLISA, Ministry of Health (MOH), Ministry of Education and Training (MOET), etc. In addition to safety laws, there are many safety-related decrees, circulars, and decisions that construction firms must comply with. As a result, firms are embarrassed and cannot apply right safety provisions when developing project safety guides as well as enforcing safety legal during construction. Thus, there is a need to investigate a construction safety legislation hierarchy, which links with state governance hierarchy of Vietnam in order to assist construction firms in their safety enforcement.

The compliance with safety laws and legislation is important to improve construction safety performance (Raheem and Hinze, 2014). However, Vietnam construction firms have not given attention to safety compliance (MOLISA, 2018). In fact, construction firms usually feedback to the Government that safety laws are unclear and it is difficult to comply all safety regulations at jobsites. Despite Government efforts to complete construction safety legislation, safety laws and regulations are still inadequate. In fact, safety compliance varies between provinces, construction firms and projects. Lack of safety compliance leads to poor safety performance at workplaces (Pham, 2018b). As a result, the amount of people died in construction industry is 19.7% of total fatalities in 2017 (MOLISA, 2018). Therefore, compliance with safety legislation plays a crucial role in reducing the rate of construction accidents in Vietnam.

In response to this status quo, this chapter consists of three objectives, following four key steps of research roadmap (Figure 5.1). First, the purpose of this chapter is to focus on a thorough review of construction safety in Vietnam from 2013 to 2017 (Section 5.3). Second, this chapter aims to develop a legal framework in order to provide a comprehensive viewpoint of construction safety legislation hierarchy linked with state governance hierarchy of Vietnam (Section 5.4). After that, the third objective of this chapter is to investigate safety compliance in Vietnam (Section 5.5) through reviewing annual Government reports and carrying out a case study of the SC Vivo Project, which is evaluated as a good example of safety compliance.

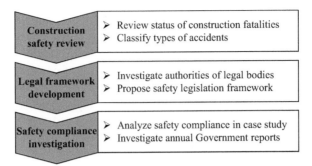

Figure 5.1 Research roadmap.

Finally, discussions (Section 5.6) are implemented to point out the limitations and solutions for safety improvement.

5.2 Research Methodology

For the first objective of chapter, authors collect a hundred annual labour accident reports of both central and provincial Governments for a decade from 2007 to 2017. However, due to a lack of data in annual reports, only reliable data from 2013 to 2017 are summarised for construction accident analysis.

For the second objective, this chapter investigates the roles, responsibilities and relationships of state management agencies including legislative, executive and judicial bodies, which are accountable for construction safety management. Through this, study develops legal framework, providing construction safety governance hierarchy and safety legal documentation hierarchy linked with firms in Vietnam construction industry.

For the third objective, due to a lack of construction safety laws, this chapter first reviews all laws, decrees, circulars and decisions related to labour safety in general. After that, the authors classify these laws and regulations by focusing on legal documents and articles specialized in construction safety. Based on these legal materials, this chapter investigates the compliance with safety laws and regulations in practice. Since safety compliance practices vary between projects, a case study is carried out in order to address in detail compliance and non-compliance issues of a real construction project, which is evaluated as a good example of safety compliance in practice. Furthermore, this chapter reviews annual Government accident reports from 2013 to 2017 to evaluate status of safety compliance in Vietnam.

5.3 Review of Construction Safety in Vietnam

Despite the important economic contribution, construction is acknowledged as a dangerous and hazardous industry due to its complex and changeable natures (Pedro, 2018). As illustrated in Figure 5.2, construction accounts for 20–30% of accidents and fatalities, which is the highest rate compared to other industries and services currently in Vietnam (MOLISA, 2018). In fact, construction accident records in Vietnam are poor because many construction firms have not reported to Vietnam government to avoid penalty, prosecution or judgement (MOLISA, 2018). Furthermore, despite the government's attention given to safety, a large number of near-miss accidents and common injuries have not recorded comprehensively by construction stakeholders. Lack of reporting construction injuries is a common safety practice in developing countries, and companies usually pay compensation (a small amount of cash) for any injuries happened to employees (Koehn, 2000). Therefore, the real accidents can be dramatically higher than the above annual statistical reports of MOLISA.

Regarding accident types, the Occupational Safety and Health Administration (OSHA) in USA defines the "Fatal Four" as the highest rate of construction accidents through years (OSHA, 2018). Similarly, as depicted in Figure 5.3, Vietnam "fatal four" consist of fall, struck-by, caught-between and electrocution, which are the top four highest rate of accidents. Despite fluctuation, fall accidents happen most at the jobsites (approximate 30%),

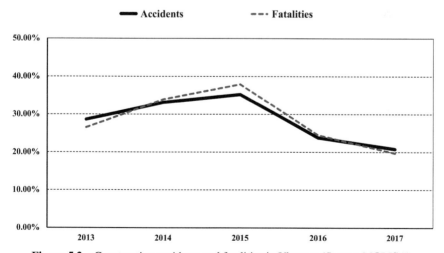

Figure 5.2 Construction accidents and fatalities in Vietnam (*Source*: MOLISA).

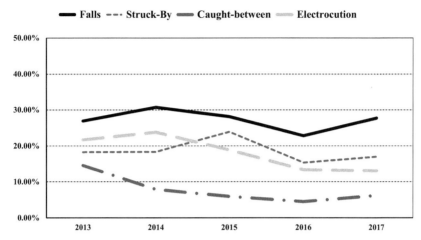

Figure 5.3 OSHA-based fatal four accidents in Vietnam (*Source*: MOLISA).

while caught-between is the smallest rate (around 10%) of occurrence among "fatal four". To significantly reduce accidents, government and construction firms therefore have to consider great concerns on analysis of these "fatal four".

In fact, the high number of accidents has caused many problems involving not only cost overruns and time delays, but also human losses and the detrimental social welfare effects (Pham, 2018b). In recent years, the Vietnam government therefore has attached great importance to completion the construction safety legal documents, as well as the supervision of safety work at construction jobsites. Particularly, there are more than 70 legal documents for labour protection and labour safety established by Vietnam government in recent 5 years. Regarding these legislation materials, more than 20 documents are directly related to construction safety. Many construction safety standards are established to specify the minimum acceptable level of safe work performance. With government efforts to reduce accidents, the number of construction fatalities is reducing significantly as shown in Figure 5.2, however, fatalities rate in construction industry is still very high. Many research points out that regulatory interpretations, understanding, and compliance plays an important role in preventing construction incidents and fatalities (Reese and Eidson 2006). Furthermore, annual safety reports of government agencies and construction associations in Vietnam have revealed that one of the root causes of accidents is a lack of compliance with legal laws and regulations. Besides a less attention paid to safety, many construction firms state that legislation is too confused and difficult to implement properly in practice. As a result, poor safety standards and law enforcement make workers vulnerable to accidents. Thus, there is a need to carry out an investigation into compliance with construction safety laws and regulations in Vietnam to exposure lessons learned for promoting safety not only in Vietnam but also in developing countries.

5.4 Construction Safety Legal Framework

Vietnam construction safety legislation is unclear to comply; this chapter therefore aims to propose the construction safety legal framework, consisting of two main hierarchies: (1) Construction safety governance hierarchy illustrating the relationship among state bodies and construction subjects; and (2) Safety legal documentation hierarchy depicting levels of laws and regulations for construction safety in Vietnam. As depicted in Figure 5.4, state legislative body is responsible for promulgating constitution and laws, while state executive bodies including central and provincial levels are accountable for establising government's decrees, ministry's circulars, safety-supported decisions, as well as codes and standards in order to assist safety implementation. All these safety legislation materials and standards are legal basis

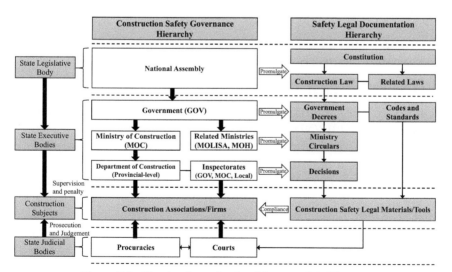

Figure 5.4 Construction safety legal framework in Vietnam.

for construction subjects (firms, managers, workers, etc.) to enforce safety. Futhermore, this executive body (state provincial level and inspectorates) plays a key role in supervising the compliance of construction subjects with safety laws and regulations in practice. State judicial bodies (procuracies and courts) would also prosecute and judge cases, which are serious injuries or fatalities based on all safety legal basic issued by these state legislative and executive bodies as well as project's contracts and specifications. The next sections would explain these hierarchies in detail.

5.4.1 State Governance Hierarchy

5.4.1.1 State legislative body

National Assembly is the state legislative body of Vietnam, which is "the highest organ of state power" to promulgate the constitution 2013, construction law (No.10/2012/QH13) and related laws (Labour Law No. 10/2012/QH13 and Law on Protection of People's Health No. 21-LCT/HĐNN8). These constitution and laws are the highest level of legal basic, which have specific provisions and articles requiring construction safety practices. In other words, safety legislation documents in the lower levels such as decrees, circulars, decisions, etc. must follow these laws promulgated by National Assembly.

5.4.1.2 State executive body

Government is state executive body, consisting of central and local governments. Central government includes government office, ministries and ministry-level agencies, while local government consists of people's committee and related departments. Upon proposals of ministry, central government will promulgate decrees for state governance in a specific field that ministry governs. For example, in terms of safety, Government Decrees (e.g., Decree No. 45/2013/ND-CP, Decree No. 110/2002/ND-CP, Decree No. 06/1995/ND-CP) are issued by Prime Minister to provide detailed regulations on the implementation of safety-related articles of the laws.

Ministries belong to central government and play a role in assisting central and provincial governments in specialized management. For example, the MOC is a government ministry responsible for state management on construction. To do this, minister issues safety-related circulars (or joint circulars with related ministries) in order to guide the implementation of the government's decrees as well as detail and guide the implementation of a number of articles of a specific law.

Local government is accountable for state administration system in provinces. Department of construction, which belongs to the people's committee (provincial-level), directly supervises construction safety activities. Based on construction laws and government's decrees, these state local organizations issue safety-related decisions within their own authorities, and these decisions only take effect on their own province.

The key ministries and ministry-level agencies responsible for construction safety are detailed as follows:

Ministry of Construction (MOC)

The MOC is a key ministry accountable for state management on all construction aspects including building materials, housing and office buildings, architecture, urban and rural construction planning, etc. Especially, construction Activities Management Agency is under the MOC, which plays a key role in advising and assisting the Minister in state governance. Furthermore, this agency implements the law enforcement in terms of construction activities such as formulation, appraisal, approval and management of investment projects on construction, building permits, building surveying, building design, construction, as well as safety and occupational health in construction activities (MOC, 2018). In other words, all construction safety are assisted and supervised by MOC.

Ministry of Labour, Invalids and Social Affairs (MOLISA)

According to decree No. 14/2017/ND-CP of Vietnam government, MOLISA performs state management function on labour, wage and salary, employment, vocational education, social insurances, occupational safety and hygiene, etc. This ministry takes care of safety issues relating to employees (workers). Especially, the Department of Work Safety (DWS), which is a unit of MOLISA, plays a key role in assisting the Minister to perform the state management function in work safety area throughout the country, as specified by laws.

Some key duties of DWS are detailed based on a MOLISA decision No. 1128/QD-LDTBXH as follows (MOLISA):

- Providing guidelines to and monitoring the implementation of State's and Ministry's regulations as the responsibilities assigned.
- Monitoring, reviewing, providing information of occupational safety and hygiene; occupational safety and hygiene statistic in accordance with regulation of law on statistic; building, managing the database on occupational safety and hygiene overall country scale.

- Conducting update, statistic, report on occupational accident; implementing the collection, archive, review, providing, announcement, evaluation on occupational accidents, occupational illness and diseases, serious technical incidents causing occupational safety – hygiene failures.
- Organizing activities of communication, information and education on occupational safety and hygiene; prevention from the technical incidents causing occupational safety and hygiene failure, occupational accidents, occupational diseases and illnesses as assigned by the Ministry.
- Providing recommendation of merits and rewards good safety performance as assigned by the Ministry.
- Taking part in research works, in training activities to specialized personnel, civil and public servants, workers in the area of work safety as assigned by the Ministry.
- Collecting information and data, making periodical and incidental reports on work safety area as the responsibilities scale of the Ministry.
- Coordination with Ministry of Health in promulgating the list of occupational diseases.
- Giving training course about occupation safety and health, labour protection.

Ministry of Health (MOH)

MOH is responsible for the state governance and guidance of the health, healthcare and health industry in Vietnam. In other words, the issues relating to health of the employees is taken care of by the MOH. Under this ministry, there are several departments and agencies involved in terms of occupational health such as Deperment of Occupation Health and Safety and Injury. The responsibilities of these organizations are as follows:

- Setting up the strategy, planning, plan for monitoring working environment, taking care of employee's health, prevention occupation disease and accident.
- Elaboration, amendment, adjustment legal documents, norms, standards and procedures for working environment, occupation disease and prevention occupation disease and accident.
- Leading, guiding, supervision of working environment, prevention and treatment occupation disease, injury.
- Guidance on implementation of periodic health check, identifying occupational disease, assesses the impact of the working environment on workers' health.

– Leading and coordinating with other relevant agencies in monitoring, inspection in workplace health care and prevention of occupational incidents.

Ministry of Science and Technology (MOST)

The MOST is a government ministry in order to performs functions of State management on science and technology. For construction safety, MOST is responsible for managing research and application of science and technology on labour safety and hygiene. Furthermore, MOST collaborates with the MOLISA and the MOH to develop, issue and manage technical standards on labour safety and specifications of personal protective devices.

Ministry of Education and Training (MOET)

The MOET is the governmental ministry responsible for the state governance in terms of general/academic education and higher education in Vietnam. Regarding construction safety, MOET duties are to direct and incorporate labour safety and labour safety into the curricula of universities, colleges, construction-specialized schools.

State Inspectorate on Occupational Safety and Health

State inspectorate functions are to inspect the compliance with safety law and regulations; investigate occupational accidents and standard violations on labour safety and carry out complaints and denunciations of labourers about violations of construction and labour laws.

In conclusion, the state executive bodies play an significant role in guiding and supervising the safety law compliance of construction subjects. Particularly, MOC and department of construction (provincial level) has main duties and responsibilities for construction safety in conjuction with related ministries such as MOLISA and MOH. In case of accidents happened, the penalties are given by local government or state inpectorates.

5.4.1.3 State judicial body

The state judicial bodies (procuracies and courts) carry out the prosecutions and judgement of serious cases relating to fatalities or serious injuries happened at construction jobsites. The system of people's courts includes the Supreme People's Courts, provincial courts, and district courts. Most cases are settled by one professional judge and two lay assessors. Judgement is based on safety legal basis promulgated by state legislative and executive bodies. Futhermore, courts refer additional materials such as construction contract signed by stakeholders and specifications of projetcs to make final judgement decision.

5.4.2 Safety Legal Documentation Hierarchy

As depicted in Figure 5.4, the safety law and regulations in Vietnam consists of levels, which are promulgated by different state governments. For example, laws, government decrees and ministry circulars are issued by national assembly, central government and ministry, respectively. Due to a lack of legal specialized for construction safety, Table 5.1 focuses on providing a list of safety laws and regulations, which took effect on construction safety:

Table 5.1   List of construction safety laws and regulations

No.	Name	Promulgated by	Date of Issuance	Brief Description	Key Safety-Related Articles
I			SAFETY-RELATED LAWS		
1	Construction Law No. 16/2003/QH11	National Assembly	26/11/2003	Promulgating the construction activities; the rights and obligations of organizations and individuals investing in the construction of works and engaging with construction activities.	Article 45 (Chapter 3), article 85 (Chapter 5), and article 87 (Chapter 5)
2	Labour Law No. 10/2012/QH13	National Assembly	18/06/2012	Chapter 9 of this Law has provisions regarding the occupation health and safety sanitation, special conditions, etc.	From article 133 to article 152 (Chapter 9)
3	Law on Protection of People's Health	National Assembly	30/6/1989	Provisions on health care for citizens	Article 14 (Chapter 2)
II			GOVERNMENT DECREES		
4	Decree No. 45/2013/ND-CP	GOV	10/05/2013	Elaborating a number of articles of the labour law on hours of work, hours of rest, occupational safety and occupational hygiene	From article 9 to article 24 (Chapter 3)

Table 5.1 (Continued)

No.	Name	Promulgated by	Date of Issuance	Brief Description	Key Safety-Related Articles
III			MINISTRY CIRCULARS AND DECISIONS		
5	Circular No. 22/2010/TT-BXD	MOC	03/12/2010	Provisions for work safety in the construction activities	Article 3 and article 4 (Chapter 2) and Chapter 3
6	Circular No. 04/2014/TT-BLDTBXH	MOLISA	12/02/2014	Pursuant to Article 149 of the Labour law, On amendment and supplement of a number of provisions of Circular No. 37/2005/TT-BLDTBXH dated 29/12/2005 guiding on occupational safety and health training	From article 4 to article 7 (Chapter 2) and article 8 (Chapter 3)
7	Circular No. 27/2013/TT-BLDTBXH	MOLISA	18/10/2013	Pursuant to Article 150 of the Labour law, providing for work safety and hygiene training	From article 4 to article 9 (Chapter 2) and Chapter 4
8	Circular No. 37/2005/TT-BLĐTBXH	MOLISA	29/12/2005	Guiding the training of work safety, workplace hygiene, etc.	From article 1 to article 15
9	Circular No. 41/2011/TT-BLDTBXH	MOLISA	28/12/2011	Amendment of the Circular No. 37/2005/TT-BL TBXH on training on work safety, workplace hygiene, etc.	
10	Joint Circular No. 01/2011/TT-BLDTBXH – BYT	MOLISA and MOH	10/01/2011	Guiding the implementation of work safety protection, sanitation and hygiene in the workplace.	From article 4 to article 6 (Chapter 2) and Chapter 3

(Continued)

Table 5.1   (Continued)

No.	Name	Promulgated by	Date of Issuance	Brief Description	Key Safety-Related Articles
11	Circular No. 19/2011/TT-BYT	MOH	06/06/2011	Provisions on management of work safety, hygiene and sanitation environment, health care for workers and prevention of occupational Diseases	
12	Decision No. 3733/2002/ QD-BYT	MOH	10/10/2002	Promulgating 21 standards of labour hygiene, 05 principles and 07 parameters of labour hygiene and occupational health	
13	Circular No. 10/1998/TT-BLDTBXH	MOLISA	28/05/1998	Provision of protective equipment for employees at the working place	
14	Decision 68/2008/QD-LĐTBXH	MOLISA	29/12/2008	Promulgating the list of personal protection for people working in dangerous conditions.	
IV			CODES AND STANDARDS		
1	Vietnam Building Code QCVN05: 2009/BXD for House and Public Buildings	MOC	2009	Provisions on occupational health and safety	
2	TCVN 3146-1986	MOC	1986	Electric welding – General requirements of safety	
3	TCVN 5041 – 89 (ISO 7731-1986)	MOC	1989	Danger alarm in the workplace – Sonic signal for danger	
4	TCVN 5308-91	MOC	1991	Technical norms for safety in construction	

Table 5.1 (Continued)

No.	Name	Promulgated by	Date of Issuance	Brief Description	Key Safety-Related Articles
5	TCVN 3288-1979	MOC	1979	Ventilation system – General requirements for safety	
6	TCVN 5744-1993	MOC	1993	Elevator – Safety requirements for installation and usage	
7	TCVN 5866-1995	MOC	1995	Elevator: Mechanical safety structure	
8	TCVN 2287-78	MOLISA	1978	Work safety standard system – Basic regulations	

5.5 Investigation of Construction Safety Compliance

5.5.1 Case Study Analysis

To investigate the compliance with safety laws and regulations in practice, a case study of "SC Vivo project" is carried out, addressing compliance and non-compliance issues. This is South Saigon complex commercial development project, comprising commercial retail, offices, serviced apartments (see Figure 5.5). The project has upon 44,373 m^2 of land and locates in Ho Chi Minh City, Vietnam. Phu Hung Gia Construction and Investment JSC (PHG) is responsible for enforcing construction safety laws and regulations in this project.

Based on characteristics of the project, PHG reviews safety laws and regulations (Table 5.1) to comply only 12 legal documents as follows:

– Construction Law No. 16/2003/QH11
– Labour Law No. 10/QH13/2012
– Circular No. 22/2010/TT-BXD
– Standard TCVN 5308:1191
– Decree No. 45/2012/ND-CP
– Circular No. 01/2011/TTLT-BLĐTBXH-BYT
– Circular No. 32/2011/TT-BL   TBXH
– Circular No. 10/1998/TT-BL   TBXH
– Circular No. 19/2011/TT-BYT
– Circular No. 37/2005/TT-BL   TBXH
– Circular No. 41/2011/TT-BL   TBXH
– Decision No. 68/2008/Q   -BL   TBXH

Figure 5.5   SC Vivo project.

Despite a good example of safety compliance in Vietnam, many compliance and non-compliance issues happen in safety enforcement (Appendix I).

Key reasons for these issues are as follows:

- The nature of construction jobsite is very large and changeable. Thus, it is very difficult for the PHG to meet the agreed safety requirements.
- Labour force only works in short time, and is often changeable. Moreover, costs of training and health examination for workers are very high. Thus, safety training and health examination are not complied.
- The company did not comply some regulations, which affect safety management indirectly such as reporting fatal accidents to inspectorate of MOLISA because the intervention of MOLISA can impact on the project performance.

Since safety laws and regulations are very wide and confused; PHG therefore focuses on safety legal documents that can directly impact on the safety performance in this project as follows:

- The regulations related to workplace in Labour No. 10/2012/QH13 and Decree No. 45/2012/ND-CP: testing and assessing dangerous factors at workplace in order to set up construction method for eliminating and avoiding risks

– The regulations related to using machine, equipment with strict safety requirements in Labour law No. 10/2012/QH13, Decree No. 45/2012/ND-CP and standard No. 5308-1991/BXD: PHG must be inspected machine, equipment before use and periodically inspected during the process of utilization by the safety-certified organization

– The regulations related to training for employees in Circular No. 37/2005/BLĐTBXH: Employees who work in unsafe environment must attend at training course on safety and health, do a test and achieve safety certificate before work.

– The regulations for using scaffold in standard No. 5308-1991/BXD: design construction method to use scaffold properly and comply regulations in installing and dismantling scaffold.

– The regulations for using electricity in standard No. 5308-1991/BXD and Circular No. 22/2010/TTBXD.

5.5.2 Status of Construction Safety Compliance

Safety management practices vary not only between construction firms but also among jobsites. To understand safety compliance in Vietnam, annual Government reports of occupational accidents are reviewed to analyze key limitations in construction safety compliance in Vietnam.

5.5.2.1 Limitations in enforcing safety laws and regulations

• Limitations at the construction firms

– There is no safety-specialized unit/department responsible for occupational safety and health in most small and medium construction enterprises (SME) as well as in few large enterprises.

– The number of full-time safety managers/supervisors in charge of construction safety has been limited. Instead, part-time safety officers from other departments lack their professional ability to enforce effectively the safety law and regulations.

– Lack of training for workers on construction safety.

– The periodical health check for workers is not paid much attention to, especially for workers signing short-term working contracts.

– Lack of compliance with regulations for reporting construction incidents and accidents comprehensively and timely.

- Limitations at the construction jobsites

    - Lack or no engineer in charge of safety at construction jobsite. In fact, this duty is done by the part-time safety manager or technician.
    - Workers are not trained on safety knowledge and skills before and during execution.
    - The safety inspection of equipment and tools used at the construction jobsite has not been carried out periodically and in accordance with the right safety process.
    - Equipment and tools for workers are not used properly, while type and quality of these equipment are not heterogeneous.
    - Lack of education and training on construction safety.

5.5.2.2 Causes of lacking of safety compliance
- Safety legislation system

    - Construction safety and labour safety has been overlapped by many levels from the central government to provincial authorities, among many ministries (specialized and non-specialized ministry such as MOC, MOLISA, MOH).
    - The system of safety legal documents in general and construction safety in particular is quite large, promulgated by many state bodies (government, ministries and provincial localities). Therefore, it is difficult to enforce construction safety.
    - Some regulations are old, not suitable with reality; thus, it is difficult to apply, especially for small businesses.

- Awareness of stakeholders involving in construction activities

    - The owners and supervisors did not pay much attention to construction safety. They state that these duties and responsibilities of contractors and workers, and thus neglect the safety compliance.
    - There are many contractors who have not considered construction safety as the most important work for improving construction production. Safety costs have not been considered as investments, which bring many benefits. Firms define safety cost as production costs, and thus they have to reduce this expenditure.
    - Activities for ensuring construction safety are less important. Firms focus on improving schedule, quantity and quality.
    - The worker's awareness of compliance with construction safety is low. The psychology of making products for high income is quite

common, and the worker therefore can skip many safety steps that increase the risk of accidents.

- The role of state governance
    - The roles and responsibilities of state management agencies are unclear. Due to a lack of clear and detailed functions of agencies, the state inspection and supervision of safety compliance is not strict.
    - The state inspectorates do not inspect periodically to prevent accidents. Thus, the effects of state management are poor and passive, mainly after construction accidents happen.
    - Penalties for violating safety regulations are not high enough to deter the contractor.

5.6 Discussion

Safety compliance practices vary among construction projects worldwide (Raheem, 2011). Similarly, complying with construction safety is different between firms and projects in Vietnam. Despite Government efforts to promulgate construction safety laws and regulations, awareness of construction stakeholders (owner, contractor, designer, supervisor, specialist, etc.) plays a crucial role to enforce safety. To reduce accidents, construction firms shall intensify the construction safety compliance and attach great importance to prevention method of construction accidents, focusing on "fatal four" such as falls, electric shock, attacked by falling objects, etc. To do this, Government in central and local levels inspectorate safety compliance in both construction firms and jobsites. Many research prove that in developing countries with safety legislation system, the governing authority is often weak or non-existent (Lee, 2003). In other words, the role of state governance is very important to improve awareness of construction firms, which can promote safety performance.

Accident reports are necessary to record unsafe action and behavior of employees. However, it differs among countries (Raheem, 2014) and even lack reporting system in the developing countries (Hamalainen, 2009). Despite its significance to improve safety, reporting accidents are not paid much attention to by Vietnam construction firms. They usually provide compensation (cash amount) for any injuries to workers and try to hide accidents to avoid penalty from Government. Moreover, there is no formal regulatory framework for worker safety to easily report incidents/accidents at construction jobsites. As a result, accident statistics announced is not adequate, and evaluation of real accident situation therefore is not reliable. Lack of effective recording of the number of construction accidents and work-related diseases negatively affects collection of safety database to update safety policies as well as lesson learned to promote safety.

5.7 Conclusion

Vietnam construction is considered as the most hazardous industry, causing numerous injuries, fatalities. Despite significant contribution to economic development, construction has the highest proportion of accidents. Safety compliance and legal system play a crucial role in preventing hazards and reducing accidents at construction jobsite. With this regard, the research aims to review construction safety in Vietnam from 2013 to 2017. After that, a legal framework is developed to provide a comprehensive viewpoint of construction safety legislation hierarchy linked with state governance hierarchy in Vietnam construction industry. Finally, this chapter investigates the compliance with construction safety laws and regulations in Vietnam. A case study of SC Vivo project is analyzed to address limitations and causes during safety enforcement in a real jobsite even though this project is evaluated to have a good safety compliance in practice. Furthermore, this chapter reviews annual Government reports in order to understand safety compliance in Vietnam. Findings reveal that awareness of construction entities and workers is very important to improve safety. Accidents would be of high chance to happen when firms fail to enforce safety laws and regulations. Despite significance, reporting accidents are not paid much attention to by construction firms and government. Lack of accident reports negatively affects collection of safety database to update safety policies as well as lesson learned to promote safety. Furthermore, construction standards are very important to ensure safety. Thus, future work would expand investigation of a gap between construction standards and safety legislation in Vietnam.

Acknowledgements

We would like to thank the Quang-Vu Pham and Phu Hung Gia Construction and Investment JSC for allowing us to use their data source of SC Vivo case study for the research.

Appendix I: List of Compliance and Non-compliance of Safety Laws and Regulations in SG Vivo Project

No.	Laws and Regulations	Status Compliance	Non-Compliance	Remark
I	Chapter 9, Labour Law No. 10/QH13/2012			
1	Item 2, Article 136: National technical regulations on labour safety and hygiene			
	Build the safety rules and working procedures.	x		
2	Article 137: Ensuring labour safety and hygiene in workplace			
	Item 1: must make a safety plan	x		
	Item 2: must comply with the national technical regulations on labour safety	x		
3	Item 1, Article 138: The responsibility of employer for occupational health and safety			
	To ensure the workplace meeting the safety requirements		x	PHG cannot ensure all requirements according to Circular No. 19/2011/TT-BYT
	To ensure the conditions on labour safety	x		
	Testing and assessing the dangerous and harmful factors at workplace	x		

(*Continued*)

No.	Laws and Regulations	Status Compliance	Status Non-Compliance	Remark
	Periodical testing and maintaining the machinery, equipment, workshops and warehouses	x		
	There must be an instruction table on labour safety in the workplace	x		
	Gathering opinions of the representative organization of labour collective.	x		
4	Article 139: Safety and health representative			
	Must appoint a person performing the work of labour safety.	x		
5	Item 1, Article 140: The responsibilities of employer in dealing with breakdowns and emergency situations			
	Making a plan for handling of incidents and emergency response and organizing the exercises periodically	x		
	Being equipped with the technical and medical facilities	x		
	Immediately implementing the remedial measures or immediately ordering to cease the operation of machinery, equipment, workplace likely to cause occupational accidents and diseases.	x		

No.	Laws and Regulations	Status Compliance	Non-Compliance	Remark
6	Article 141: Allowance in kind for the employee working in dangerous and hazardous conditions			
	The person working in dangerous and hazardous conditions shall receive the allowance		x	PHG does not allowance in kind for employers. In some cases, PHG provides allowance for food or soft water such as concrete activity, etc.
7	Article 142: Labour Accident			
	Item 3: All of labour accidents and occupational disease and serious breakdown at workplace must be declared, investigated, counted, and reported according to Government regulations.		x	PHG adopted declaration, investigation, and report to State authority according to Circular No. 12/2012/TT-BLĐTBXH and these reports are made periodically.
8	Article 143: Occupational Diseases			
	The people catch occupational diseases must be treated carefully, tested health examination periodically and had health records separately.		x	PHG only organizes health examination to determine occupational diseases for long-time working

(Continued)

No.	Laws and Regulations	Status Compliance	Status Non-Compliance	Remark
				employees and they are made health records and treated carefully (about 20–30 people).
9	Article 144: The responsibilities of employer for people who catch occupational disease and accident			
	Making compensation to the employee suffering the occupational accident and disease	x		
10	Article 147: Inspection of machinery, equipment and materials with strict requirement on labour safety	x		
11	Article 148: Plan on labour safety and health	x		
12	Article 149: Personal protection equipment (PPE)	x		
13	Article 150: Training on safety and health	x		
14	Article 151: Safety and health information	x		
15	Article 152: Health care for employee			
	The employer must rely on the health standards regulated for each type of work for recruitment and arrangement of employees.		x	PHG only rely on the results of health examination form of some employees to arrange their work.

No.	Laws and Regulations	Status Compliance	Non-Compliance	Remark
	Organize periodic health examinations for the employee.	x		
	Employees work in environment that contract to occupational disease easily must test health.		x	The most of employees are free labour, PHG therefore does not organize for testing occupational disease. By contrast, the total schedule of the project is not long and the numbers of employees are usually changed, while the cost for testing health is also very high.
	The employee with occupational accident and disease must receive a medical examination.		x	PHG only pays the full cost for employees and compensates on agreement for them when they meet accident or occupational disease.

(Continued)

No.	Laws and Regulations	Status		Remark
		Compliance	Non-Compliance	
	The employee working at the place where there are toxic and infectious factors, upon the end of the working hours, the employer must guarantee the measures of decontamination and sterilization		x	PHG only equip hygiene conditions for workers in construction site. The workers must self-clean for themselves; PHG does not equip the measures of decontamina-tion and sterilization.
II	Decree No. 45/2013/ND-CP			
1	Article 10: Making plan for ensuring occupational safety and hygiene	x		
2	Article 13: Statement, investigation, statistics, report, compensation, pays for occupational accidents, occupational illness, and serious emergencies			
2.1	Statement, investigation, statistics, report for occupational accidents, occupational illness, and serious emergencies are regulated following	x		
a	Employers shall report fatal occupational accidents and serious occupational accidents that hurt at least 02 employees and serious		x	PHG only concentrate on taking care and compensating for family of worker in

No.	Laws and Regulations	Status Compliance	Non-Compliance	Remark
	accidents to Inspectors of local Services of MOLISA			accident. The report to Inspectors of local Services of MOLISA is not executed because PHG does not want to stop building and to be punished.
b	Employer must investigate incident, accident that injured employee and serious break-down	x		
c	Employer must statistic into book and report to HCM City Services of Labour, War Invalids and Social Affairs periodically 6 months.	x		
2.2	Employer must compensate, pay for employees who are suffered occupational accident and disease according to instruction of MOLISA.		x	PHG relies on the cost of treatment of hospital to pay for employee. Beside, in some cases, the company makes agreement of compensation with employee to pay.

(Continued)

No.	Laws and Regulations	Status		Remark
		Compliance	Non-Compliance	
3	Article 14: Controlling hazards	x		
4	Article 23: The responsibilities of using machine, equipment and material that have strict requirements on labour safety	x		
III	Circular No. 01/2011/TTLT-BLĐTBXH-BYT			
1	Article 4: Organization of safety and hygiene	x		
2	Article 6: The authorities of safety and hygiene department	x		
3	Article 15: Planning safety and hygiene for organization	x		
4	Article 16: Implementation of labour safety and sanitation plans	x		
5	Article 17: Self-inspection of Labour safety and sanitation	x		
6	Article 18: Statistic and Report	x		
7	Article 19: Preliminary and Final Review			
	Review the labour safety and sanitation work, covering analysis of results, shortcomings, problems and lessons; commendation of units			Safety Department of the company only summarizes implementation

No.	Laws and Regulations	Status Compliance	Non-Compliance	Remark
	and individuals that have well performed labour safety and sanitation work in the establishments; and launching of emulation movements to assure labour safety and sanitation. Preliminary and final reviews shall be conducted from workshops and production teams to labour-employing establishments.		x	of safety and hygiene plan and reports it to General Director in the final year. This is because, the beginning construction projects is not fixed in the instance time. Moreover, project schedule is often short.
IV	Circular No. 32/2011/TT-BLĐTBXH			
1	Article 3: The responsibilities of organization that use inspection object	x		
V	Circular No. 10/1998/TT-BL TBXH			
			x	PHG only instruct some key PPEs such as safety belt, etc. PHG only check PPE with strict requirement on safety at the first time. PHG only rely on specifications of

(*Continued*)

No.	Laws and Regulations	Status		Remark
		Compliance	Non-Compliance	
				manufacture, State certificate of quality
VI	Decision No. 68/2008/QĐ-BL TBXH			
		x		
VII	Circular No. 19/2011/TT-BYT			
1	Article 4: The content of hygiene management			
	Make record for occupational hygiene	x		
2	Article 5: The content of occupational health management			
2.1	Testing, classifying before recruiting according to Appendix No. 02, circular No. 13/2007/TT-BYT in 21/11/2007 of MOH-regulate health examination and arrangement of task for employee		x	PHG bases on health certificate of employees recognized by MOLISA to recruit. However, these health certificates do not show dangerous diseases such as heart diseases, blood pressure diseases, etc.
2.2	Testing health periodically		x	PHG do not perform health test for employee periodically. PHG just organize health test for employee

No.	Laws and Regulations	Status		Remark
		Compliance	Non-Compliance	
				working at the company for a long time.
2.3	Testing occupational disease		x	PHG does not test occupational disease because the schedule of project is not so long and labour force is always changed.
2.4	First aid for labour accident	x		
VIII	Circular No. 37/2005/TT-BLĐTBXH			
1	The content of training	x		
2	Item 2, section II: Training for employee	x		
3	Section IV: Training for safety and hygiene staff	x		
4	Section VI: Labour safety and hygiene certificate and safety card.			
	A certificate of occupational safety and health training attendance shall be issued to employers and occupational safety and health officials after they meet the requirements of training tests	x		
IX	Circular No. 41/2011/TT-BLĐTBXH			
1	Employers can print and manage safety card according to Appendix			

(Continued)

No.	Laws and Regulations	Status Compliance	Status Non-Compliance	Remark
	No. 01 of this circular.	x		
X	Construction Law No. 16/2003/QH11			
1	Article 73: Building conditions: Contractors have equipment for execution of building works, which guarantees safety and quality of the works	x		
2	Article 78: Safety in building	x		
XI	Circular No. 22/2010/TT-BXD			
1	Article 3: General Requirements on construction sites	x		
2	Article 4: Requirements during construction			
	Workers at the construction site are provided with medical checks-up and safety training and adequate personal safety equipment under the labour law		x	PHG does not check worker's health because they work in short time. PHG just check health for workers who work for long time.
3	Article 6: The responsibility of construction contractor			
	Assume the prime responsibility for and coordinate with investors in overcoming consequences,			When accident happens, PHG solves these problems in internal

No.	Laws and Regulations	Status Compliance	Status Non-Compliance	Remark
	declaring, investigating, and making records on incidents or labour accidents at the construction site		x	company. In detail, PHG reaches an agreement in compensation with employees.
4	Article 10: The responsibility of contractor's safety staff	x		
5	Article 11: Handling of incidents when labour accidents occur			
	The investor, contractor and concerned units shall promptly report the accident to relevant management agencies for examination and inspection under regulations to identify causes of the incident and accident		x	Just adopting compensation and treatment for employees when they get injuries. Reaching an internal compensation agreement with employees.
XII	Standard TCVN 5308:1191			
1	Section 1: General regulations	x		
2	Section 2: Organization of overall plan	x		
3	Section 3: Install and use electricity in construction	x		
4	Section 6: Use construction machines			

(*Continued*)

No.	Laws and Regulations	Status		Remark
		Compliance	Non-Compliance	
	Construction machines will maintain technique and periodically repair under regulations in technique documents.		x	PHG depends on technical documents of machine to maintain periodically. Equipment department only find out same parts to replace them without calculated design.
	Structure of construction machines must be guaranteed for warning sign in dangerous situations		x	PHG cannot equip modern machines that have warning sign in abnormal situations. Since construction machines are often old and outdate, the company cannot ensure for this requirement.
5	Section 8: Install, usage, uninstall scaffold, board			
	Types of scaffold and scaffold board are approved by Authority.	x		

No.	Laws and Regulations	Status		Remark
		Compliance	Non-Compliance	
	When scaffold is over 6 m high, there shall be at least two working floors.		x	Due to lack of scaffold board, site management cannot provide two working floors.
	The width of working floor of scaffold and scaffold board is at least 1 m. When transporting material on working floor by using trolley, this width of floor is at least 1.5 m.		x	Due to lack of scaffold boards
	Scaffold boards with a height of 4 m are only used after leader of group acceptance. More than 4 m high scaffolds are inspected by Technical department.		x	All scaffolds 4 m high or above are permitted for usage when leader of group accepted.
	Every day before working, technical staff of construction site or leader of group must inspect scaffold and board conditions.		x	Due to lack of engineers for inspecting every day, the company just inspects scaffold one time per month for scaffolds supporting concrete work, play cover.

(*Continued*)

No.	Laws and Regulations	Status		Remark
		Compliance	Non-Compliance	
	After stopping working on scaffold, board over one month, they are to be approved again for re-use.	x		
	Compliance of suitable process and follow design instructions of uninstalling scaffold.	x		

References

Bureau of Labor Statistics, Injury Report (2017). https://www.bls.gov/iif/oshcfoi1.htm. (accessed March 24, 2018).

BLS, OSHA: Commonly Used Statistics, Osha.Gov. (2016). https://stats.bls.gov/iif/oshcdnew.htm (accessed December 22, 2017).

Gambatese, J. A., M. Behm and S. Rajendran (2008). "Design's role in construction accident causality and prevention: Perspectives from an expert panel." Safety science 46(4): 675–691.

Hämäläinen, P., Saarela, K., Takala, J., (2009). Global trend according to estimated number of occupational accidents and fatal work-related diseases at region and country level. J. Saf. Res. 40, 125–139.

Hussain, R., A. Pedro, D. Y. Lee, H. C. Pham and C. S. Park (2018). "Impact of safety training and interventions on training-transfer: targeting migrant construction workers." International journal of occupational safety and ergonomics: 1–13.

Koehn, E., A. A. Sheikh and S. Jayanti (2000). "Variation in construction productivity: developing countries." AACE International Transactions: I4A.

Lingard, H. (2013) Occupational health and safety in the construction industry, Construction Management and Economics. 31, 505–514. doi:10.1080/01446193.2013.816435.

Le, Q. T., Pedro, A., Pham, H. C., and Park, C. S. (2016). A Virtual World Based Construction Defect Game for Interactive and Experiential Learning. International Journal of Engineering Education, 32(1), 457–467.

Le, Q. T., Lee, D. Y., and Park, C. S. (2014). A social network system for sharing construction safety and health knowledge. Automation in Construction, 46, 30–37.

Lee, S., Halpin, D.W., 2003. Predictive tools for estimating accident risk. J. Constr. Eng. Manage. 129(4), 431–436.

MOC, M. O. C. (2018). "Construction Activities Management Agency", from http://www.moc.gov.vn/introduction/-/tin-chi-tiet/7cG4/148/99833/construction-activities-management-agency.html.

MOLISA (The Ministry of Labour, I. A. S. A. (2018). "Vietnam labour accident report in 2017." Retrieved September, 2018, from http://antoanlaodong.gov.vn/catld/pages/chitiettin.aspx?IDNews=2148.

MOLISA, M. O. L.-I. A. S. A. (2018). "Department of Work Safety." September, 2018, from http://www.molisa.gov.vn/en/Pages/Detail-organization.aspx?tochucID=12.

Murie, F. (2007). "Building safety—An international perspective." International journal of occupational and environmental health 13(1): 5–11.

OSHA, O. S. a. H. A. (2018). "United States Department of Labor." Retrieved August, 2018, from https://www.osha.gov/oshstats/commonstats.html.

Perspectives, G. C. and O. Economics (2013). "Global Construction 2025. A global forecast for the construction industry to 2025." Global Construction Perspectives and Oxford Economics: 28–325.

Pedro, A., Pham, H. C., Kim, J. U., and Park, C. (2018). Development and Evaluation of Context-based Assessment System for Visualization-Enhanced Construction Safety Education. International Journal of Occupational Safety and Ergonomics, (just-accepted), 1–33.

Pedro, A., Le, Q. T., and Park, C. S. (2015). Framework for integrating safety into construction methods education through interactive virtual reality. Journal of Professional Issues in Engineering Education and Practice, 142(2), 04015011.

Pham, H. C., N.-N. Dao, J.-U. Kim, S. Cho and C.-S. Park (2018a). "Energy-Efficient Learning System Using Web-Based Panoramic Virtual Photoreality for Interactive Construction Safety Education." Sustainability 10(7): 2262.

Pham, H. C., N.-N. Dao, A. Pedro, Q. T. Le, R. Hussain, S. Cho and C. S. Park (2018b). "Virtual Field Trip for Mobile Construction Safety Education Using 360-Degree Panoramic Virtual Reality." International Journal of Engineering Education 34(4): 1174–1191.

Pham, H. C., A. Pedro, Q. T. Le, D.-Y. Lee and C.-S. Park (2017). "Interactive safety education using building anatomy modelling." Universal Access in the Information Society: 1–17.

Raheem, A., Hinze, J., Azhar, S., Choudhry, R., Riaz, Z., 2011. Comparative analysis of construction safety in Asian developing countries. In: Sixth International Conference on Construction in the 21st Century (CITC-VI), "Construction Challenges in the New Decade", Kuala Lumpur, Malaysia, pp. 623–630.

Raheem, A. A. and J. W. Hinze (2014). "Disparity between construction safety standards: A global analysis." Safety Science 70(Supplement C): 276–287.

Reese, C. D. and J. V. Eidson (2006). Handbook of OSHA construction safety and health, Crc Press.

Rowlinson, S. (2004). Construction safety management systems, Routledge.

Van Luu, T., S.-Y. Kim, T.-Q. Truong and S. O. Ogunlana (2009). "Quality improvement of apartment projects using fuzzy-QFD approach: A case study in Vietnam." KSCE Journal of civil engineering 13(5): 305–315.

Zaira, M. M., and Hadikusumo, B. H. (2017). Structural equation model of integrated safety intervention practices affecting the safety behaviour of workers in the construction industry. Safety science, 98, 124–135.

Index

About the Editor

João Paulo Davim received his Ph.D. in Mechanical Engineering in 1997, M.Sc. in Mechanical Engineering (materials and manufacturing processes) in 1991, Mechanical Engineering degree (5 years) in 1986 from the University of Porto (FEUP), the Aggregate title (Full Habilitation) from the University of Coimbra in 2005, and the D.Sc. from London Metropolitan University in 2013. He is Senior Chartered Engineer in the Portuguese Institution of Engineers with an MBA and holds a Specialist title in Engineering and Industrial Management. He is also Eur Ing by FEANI-Brussels and Fellow (FIET) by IET-London. Currently, he is Professor at the Department of Mechanical Engineering, University of Aveiro, Portugal. He has more than 30 years of teaching and research experience in Manufacturing, Materials, Mechanical and Industrial Engineering, with special emphasis in Machining & Tribology. He has also interest in Management, Engineering Education, and Higher Education for Sustainability. He has guided a large number of postdoc, Ph.D. and master's students as well as has coordinated and participated in several financed research projects. He has received several scientific awards. He has worked as an evaluator of projects for the ERC-European Research Council and other international research agencies as well as examiner of Ph.D. thesis for many universities in different countries. He is the Editor in Chief of several international journals, Guest Editor of journals, Books Editor, Book Series Editor, and Scientific Advisory for many international journals and conferences. Presently, he is an Editorial Board member of 30 international journals and acts as a reviewer for more than 100 prestigious Web of Science journals. In addition, as an editor (and co-editor), he has also published more than 100 books and has authored (and co-authored) more than 10 books, 80 book chapters, and 400 articles in journals and conferences (more than 250 articles in journals indexed in Web of Science core collection/h-index 51 citations, SCOPUS/h-index 56+/10500+ citations, Google Scholar/h-index 72+/17000+).